技工院校专业英语教材（中/高级）

烹饪实用英语

（第四版）

Cooking English (Fourth Edition)

王泳娣　主编

中国劳动社会保障出版社

图书在版编目（CIP）数据

烹饪实用英语 / 王泳娣主编 . -- 4 版 . -- 北京：中国劳动社会保障出版社，2024. --（技工院校专业英语教材）. -- ISBN 978-7-5167-6798-6

I. TS972.1

中国国家版本馆 CIP 数据核字第 2024UT0398 号

中国劳动社会保障出版社出版发行

（北京市惠新东街 1 号　邮政编码：100029）

*

北京市艺辉印刷有限公司印刷装订　新华书店经销
787 毫米 × 1092 毫米　16 开本　9.5 印张　181 千字
2024 年 12 月第 4 版　2024 年 12 月第 1 次印刷

定价：21.00 元

营销中心电话：400-606-6496
出版社网址：https://www.class.com.cn
https://jg.class.com.cn

版权专有　　　侵权必究

如有印装差错，请与本社联系调换：（010）81211666
我社将与版权执法机关配合，大力打击盗印、销售和使用盗版图书活动，敬请广大读者协助举报，经查实将给予举报者奖励。
举报电话：（010）64954652

前　言

技工院校专业英语教材自出版以来，受到了广大师生的好评。随着我国经济的进一步发展和全球经济一体化进程的加快，越来越多的企业对技术工人的专业外语水平提出了更高的要求。专业英语已成为学生们顺利择业、就业的工具之一。为了适应这一需要，我们对专业英语教材进行了修订，并扩充开发了数个热门专业的英语教材。在编写过程中我们坚持以下原则：

第一，根据专业需要划分单元结构，突出专业中与英语紧密相关的内容。精选通俗易懂的专业材料作为专业英语教材的载体，力求收录各专业最新、最实用的词汇、用语和表达，从而使教材既具有专业特色，又充分体现英语教学规律。

第二，根据不同专业对英语教学的要求，教材设计各有侧重，如侧重口语、阅读等。

第三，教材在重点段落后均设计了形式多样、联系紧密的练习，旨在达到即学即练即会的学习效果。

第四，教材以多变的学习模块、活跃的版式、充实的图片、简洁的中文提示语，增强英语学习的趣味性和易懂性。

专业英语教材自成体系，同时每种教材的编写又参照了相关专业的教学计划和主要专业课程的教学大纲，故可与相关专业教材配套使用。本版教材均配有对话和阅读录音、重要内容翻译、电子课件等，以方便师生教与学。相关资源可从技工教育网（https：//jg.class.com.cn）下载。

　　本教材由王泳娣主编。为进一步完善本教材的编写，服务于技工院校专业英语的教学，读者可将对本教材的意见和建议发送至邮箱：ggk@class.com.cn。

2024 年 12 月

内容简介

《烹饪实用英语（第四版）》以"岗位"为中心组织理论与实操知识，对厨房每个岗位需要完成的任务进行分解，在学生熟悉的工作环境中开展专业英语的学习。

本教材包括 7 个单元。Pre-Unit 介绍烹饪基础知识，其余单元根据不同岗位的要求设置内容。

每课分为 7 个部分。Goal 展现每课的学习目标。Warm-up 以趣味小题的形式引出与本课相关的专业术语。Dialogue 以刚从学校毕业的学生 Leo 到一家餐饮企业轮岗实习为主线，以对话的形式，真实反映厨房内的交流和工作流程。Role-play 为对话习题的延伸，引导学生学以致用。Reading 为阅读短文，锻炼学生英文阅读能力。Task 为阅读短文习题的延伸，活跃课堂气氛。Further Study 中多种多样的学习资料是对每课的有益补充。

Contents

Pre-Unit
Basic Knowledge of Culinary Art / 1
烹饪基础知识

Unit 1
The Prep Cook / 9
帮厨组厨师

    Lesson 1   The Food Preparation Worker   食品加工工人 / 10
    Lesson 2   The Sauce Chef   调味厨师 / 16
    Lesson 3   The Kitchen Assistant   厨师助理 / 21

Unit 2
The Pastry Chef / 27
面点组厨师

    Lesson 1   The Western Pastry Chef   西式面点师 / 28
    Lesson 2   The Chinese Pastry Chef   中式面点师 / 33

Unit 3
The Cold Dishes Chef / 41
冷菜组厨师

    Lesson 1   The Chef Garde Manger   冷房厨师 / 42
    Lesson 2   The Glacier Chef   冷冻菜厨师 / 48

Unit 4
The Hot Dishes Chef / 55
热菜组厨师

    Lesson 1   The Fry Chef   煎炸厨师 / 56
    Lesson 2   The Saute Chef   烹炒厨师 / 61
    Lesson 3   The Grill and Roast Chef   烧烤厨师 / 66
    Lesson 4   The Vegetable Chef   蔬菜厨师 / 71

Unit 5
The Larder Chef / 77
肉类组厨师

    Lesson 1   The Butcher   肉类处理工 / 78
    Lesson 2   The Fish and Seafood Chef   海鲜厨师 / 83

Unit 6
The Executive Level / 89
行政组厨师

    Lesson 1   The Line Cook   三厨 / 90
    Lesson 2   The Sous Chef   副总厨 / 95
    Lesson 3   The Executive Chef   行政总厨 / 100

Appendix 1   Grammar in Use / 105
Appendix 2   Words and Expressions / 115
Appendix 3   Vocabulary / 121
Appendix 4   Translation / 127

Pre-Unit

Basic Knowledge of Culinary Art

烹饪基础知识

Goal

Learn the categories of tools in the workplace and posts in a kitchen.

Warm-up

Write the correct term below each picture.

| tank | kitchen | peeler | stove | chopping board |
| worktable | electric oven | refrigerator | soup pot | |

1. _____

2. _____

3. _____

4. _____

5. _____

6. _____

Pre-Unit
Basic Knowledge of Culinary Art

7. _____ 8. _____ 9. _____

Dialogue

行政主厨 Smith（S）带轮岗实习的 Leo（L）与副总厨 Sabina（Sa）、三厨 Jackie（J）等人见面。（team members=T）

S : Good morning, everyone. I have an important **announcement**. We have a new member joining our team today. Please welcome Leo!

T : Welcome, Leo!

L : Thank you, everyone. I'm really excited to be here. My name is Leo White. I **graduated** from ABC School this year.

S : Leo, I'd like you to meet some of our key team members. This is our Sous Chef, Sabina. She **oversees** the daily **operations** in the kitchen.

Sa: Hi, Leo. It's great to have you. Our kitchen is the heart of our restaurant, and it's where the magic happens. You'll find it organized and efficient.

L : Thank you, Sabina. I look forward to learning from you.

S : Next, you'll work closely with our Line Cook, Jackie. He will show you around the restaurant and the workplace.

J : Welcome, Leo. Feel free to ask me any questions about our kitchen.

L : Thank you, Jackie. I'm sure I'll have plenty of questions!

Role-play

Make a simple dialogue between Jackie and Leo on their way around. You may need the words and expressions below.

English	Chinese	English	Chinese
rough processing area	粗加工区	staple food area	主食区
cleaning area	清洗区	pastry area	面点区
operation area	操作区	meal preparation area	备餐区
cooking area	烹饪区	food storage area	食物储藏区

EXAMPLE: J: Leo, this is a pastry area in the kitchen. It is used to make noodles, steamed buns, and so on.

L: Wow! We have all kinds of baking tools here.

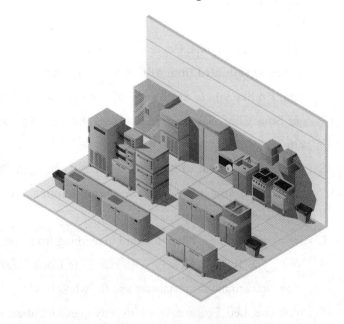

Pre-Unit
Basic Knowledge of Culinary Art

Reading

Food is a **universal** language that brings people together. Cooking is not only about **flavors** but also about **cultures** and **traditions**. Just like French fries, hamburgers, pizza, *zongzi*, mooncakes and dumplings, each dish tells a **unique** cultural and traditional story. Food is one of the best ways to connect people from all over the world.

Ever since I was a little kid, I've dreamed of becoming a chef. I enjoy making people happy through the food I prepared well. Being a chef means not only cooking but also **innovating**, **experimenting**, and sharing the joy of food with others.

Food has no boundaries, and neither does my dream.

A. Write the correct term below each picture according to the reading material.

1. _____ 2. _____ 3. _____

4. _____ 5. _____ 6. _____

B. Match the food with the corresponding country. Draw lines.

1. *zongzi*   a. Italy
2. French fries   b. Germany
3. hamburgers   c. China
4. pizza   d. Belgium

C. Learn the important phrases in the text.

1. dream of 梦想

练习：I've dreamed of becoming a chef.

中文：_____

2. no...neither 也不

● 用于倒装结构，表示前者的否定情况也同样适用于后者。

练习：Food has no boundaries, and neither does my dream.

中文：_____

Task

Work in groups. Find out as many working tools as possible that you will use in your future job. Then, present the results to the whole class. You may need the words and expressions below.

English	Chinese	English	Chinese
electric rice cooker	电饭锅	cleaver	砍骨刀
steamer	蒸锅	knife sharpener	磨刀器
frying pan	煎锅	spatula	铲子
ice making machine	制冰机	whisk	搅拌器

Further Study

Learn about posts in a kitchen.

分类	岗位名称	岗位职责
Prep Cook 帮厨组厨师	Food Preparation Worker 食品加工工人	负责食品的清洗、切剁、存放，以便厨师烹调时使用
	Sauce Chef 调味厨师	负责为各种菜肴调味
	Kitchen Assistant 厨师助理	负责保证食品质量和厨房内的清洁卫生与安全工作

续表

分类	岗位名称	岗位职责
Pastry Chef 面点组厨师	Western Pastry Chef 西式面点师	负责西式面点的制作
	Chinese Pastry Chef 中式面点师	负责中式面点的制作
Cold Dishes Chef 冷菜组厨师	Chef Garde Manger 冷房厨师	负责制作冷菜色拉、冷汤、罐装或腌制食品、调味品和盘中装饰、冰雕等
	Glacier Chef 冷冻菜厨师	冷房厨师的下属。在冷房厨师的指导下，在厨房中扮演很多角色，负责准备凉的点心或冷冻点心
Hot Dishes Chef 热菜组厨师	Fry Chef 煎炸厨师	负责煎、炸、烘烤和爆炒食物
	Saute Chef 烹炒厨师	负责烹炒食物
	Grill and Roast Chef 烧烤厨师	负责烧烤食物
	Vegetable Chef 蔬菜厨师	负责炒、炖、煮或做各种蔬菜冷盘，如蔬菜色拉和蔬菜寿司等
Larder Chef 肉类组厨师	Butcher 肉类处理工	负责屠宰动物、切割肉类
	Fish and Seafood Chef 海鲜厨师	烹饪各种水产品和海产品
Executive Level 行政组厨师	Executive Chef 行政总厨	监管所有厨房工作人员、食品加工或烹制
	Sous Chef 副总厨	处理各项事务，确保厨房食品的质量
	Line Cook 三厨	掌管不同的厨房小组，是各组组长。向行政副总厨直接汇报

Unit 1

The Prep Cook

帮厨组厨师

Lesson 1
The Food Preparation Worker
食品加工工人

Goal

Learn how to store, pack and prepare foods.

Warm-up

A. Write the correct term below each picture.

disposable food container	thermo food container
food container with locks	salad-and-go food container
fresh sealed food container	paper food container

1. _____

2. _____

3. _____

Unit 1
The Prep Cook

4. _____ 5. _____ 6. _____

B. Look at the following pictures and tell the name of each food tagging in Chinese.

1. date tagging   2. place of origin   3. storage conditions   4. instructions for use

_____   _____   _____   _____

Dialogue

Betty（B）是餐厅的食品加工工人，她在教 Leo（L）做腌肉。

L: Betty, the pickled meat tastes so nice. I want to learn how to make it.

B: OK. Actually it is very simple. Do you have a stew pot?

L: Yes. Is this OK?

B: No, this is too small. We need a large one.

L: OK, how about this one?

B: Perfect. And we also need a large **crockery** or a glass container.

L: I see. Here it is.

B: Great. Now let's get 6 pounds of salt, 1 pound of sugar, and 4 gallons of water.

L: Here, 6 pounds of salt, 1 pound of sugar and 4 gallons of water. What should we do next?

B: Now we bring 4 gallons of water to the boil over high heat. OK. It is at a rolling boil. Now add all of the salt and sugar.

L: Oh, I see.

B: Remove the pot from the fire and cool the pickling mixture to room temperature.

L: And then?

B: Then we pour the cooled pickling mixture into the large crockery and add the meat. Remember to place a clean cutting board on top of it to keep the meat totally under the pickling **liquid**.

L: But how can I do it?

B: I usually use a clean heavy flat stone.

L: Is it really so simple? I just can't wait to taste it.

B: No, you need to leave the meat in the pickling liquid for three days.

L: You mean three days later I will be able to enjoy the wonderful pickled meat? Oh, great! Thank you so much for teaching me.

B: My pleasure. Oh, by the way, you can keep the mixture for reuse.

A. Reorder the letters.

Letters	Words
i p k c l e	p
ll ga on	g
ner tai con	c
i i qu d l	l

B. Choose the ingredients for the pickled meat according to the dialogue.

Ingredients: _____

a. salt

b. butter

c. water

d. chili

Unit 1
The Prep Cook

e. sugar   f. soy sauce   g. vinegar   h. yogurt

Role-play

Make a simple dialogue between Betty and Leo about how to make the pickled meat.

Reading

There are so many **recipes** that call for **diced**, **sliced** and **chopped onions**.[1] Think about how many times you cut onions each week and you will understand why it is so important you learn how to do it safely and properly.[2]

Using a chef's knife, cut the **stem** end almost off but leave a little to **grab**, so you can start peeling. Peel all the outside skin off.

Place the onion on the cutting board with the **root** end facing up, and slice the onion in half long ways. By leaving the root **attached**[3], it will help keep the onion together while slicing.

Take each half of the onion and lay it down flat on your cutting board. Make **multiple** cuts long ways from top to bottom but not through the root at the end.

Turn the onion 90 **degrees** and make multiple cuts across the onion. Being sure to keep your **fingers** curled under, so you wouldn't cut them. How many slices will gain depends on how fine a dice you want.

A. Match the Chinese with the English.

(   ) The onion is a little rotted at the stem end.
(   ) She cut the apple into halves, and then cut across each half once to make 4 equal shares.
(   ) When peeling an onion, you should grab its stem end.

1. 她先把那个苹果切成两半，然后再把每一半横切一刀，这样就把苹果分成了4等份。
2. 剥洋葱的时候，你应该抓住其根茎端。
3. 洋葱根茎端有点烂了。

B. Learn the important grammar and phrases in the text.

1. there be 句型
- There are so many recipes that call for diced, sliced and chopped onions. 很多菜肴都需要配洋葱丁、洋葱片或洋葱末。
- there be 句型是英语中的常用句型，意思是"有"，表示"存在"或"发生"。there 在此结构中是引导词，无实际意义。谓语动词应和其后出现的主语在单复数上一致。以此句为例，主语 recipes 是复数，所以谓语动词用 are。

练习：瓶子里有些苹果汁。
提示：apple juice
英文：_____

2. 宾语从句
- Think about how many times you cut onions each week and you will understand why it is so important you learn how to do it safely and properly. 只要想想你每周要切多少次洋葱，你就明白学会安全正确地切洋葱为何如此重要了。
- 不少动词后面可跟连接代词或连接副词引导的宾语从句。句中，how many times you cut onions each week 和 why it is so important you learn how to do it safely and properly 分别位于 think about 和 understand 之后，构成宾语从句。how many 和 why 分别是连接代词和连接副词。

练习：You'd better work out how much we will spend during the trip.
中文：_____

Unit 1
The Prep Cook

3. by doing sth. 通过做某事

- By leaving the root attached, it will help keep the onion together while slicing. 通过将根部靠拢并齐，让洋葱在切的时候不易散。

练习：他靠写作为生。

提示：earn one's living

英文：_____

Task

Group discussion: reorder the pictures according to the passage and describe the procedures for cutting onions with the useful phrases below.

Right order: _____

　　　　a.　　　　　　　　　　b.　　　　　　　　　　c.

1. place...on, slice...in half long ways, leave
2. each half of, lay...down, from top to bottom
3. turn, multiple cuts across

Further Study

Learn to be a qualified food preparation worker.

A food preparation worker works in the back of the kitchen department. He or she reports directly to the sous chef. The head chef and all auxiliary（辅助的）chefs are over him as well.

The main job of a food preparation worker is to convert（转变）the food into usable pieces. He or she must break down all foods  into the correct sizes and precook it if necessary. He or she also needs to inspect（检查）the quality. He or she must package all foods into conveniently located containers that are ready for the use of all cooking staff in a timely manner.

Lesson 2
The Sauce Chef
调味厨师

Goal

Learn the English names of various kinds of sauce and the nutrition in different kinds of soup.

Warm-up

Write the correct term below each picture.

| ham | brown sugar | mustard sauce |
| bread crumbs | egg yolk | strawberry sauce |

1. _____ 2. _____ 3. _____

Unit 1
The Prep Cook

4. _____ 5. _____ 6. _____

Dialogue

Celine（C）是餐厅的调味厨师，她在教 Leo（L）做芥末酱火腿。

L: Celine, what are you doing?

C: I'm making ham loaves with **mustard** sauce.

L: Sounds delicious. How do you make it?

C: I've combined the bread crumbs, milk and eggs in a large bowl. And now, I'll crumble some meat over the mixture and mix them well.

L: OK. What can I do for you?

C: Please help me combine brown sugar and **cloves** with mustard in a small bowl and spread the mixture on two loaf pans.

L: OK.

C: I'll press the meat mixture on top.

L: Oh, I see.

C: **Bake** the meat mixture, uncovered, at 350 ℉* for one hour until a meat **thermometer** reads 160 ℉. Let's wait for 10 minutes.

(*10 minutes later.*)

C: Let me put sauce, and combine egg **yolks**, mustard, **vinegar**, sugar, water and salt in a saucepan. Then cook and stir it over low heat until the mixture is thickened and reaches 160 ℉ about 5 minutes.

---

* 350 ℉：350 华氏度，相当于 176.7 摄氏度。英文：three hundred and fifty degrees Fahrenheit。

L: Time's up. Can we remove it from the heat?

C: Yes. Stir in butter and horseradish. Cool. Fold in cream. Serve with ham loaves.

A. Match the English with the Chinese. Draw lines.

1. bread crumbs   a. 把……与……混合
2. brown sugar   b. 时间到。
3. combine...with...   c. 面包屑
4. Time's up.   d. 红糖

B. Choose the ingredients for the ham loaves with mustard sauce according to the dialogue.

Ingredients: _____

a. brown sugar

b. mustard

c. ham

d. bread crumbs

e. strawberry sauce

f. ketchup

g. clove

h. milk

i. onion

j. egg

Role-play

Make a simple dialogue between Celine and Leo about how to make the ham loaves with mustard sause.

Reading

Soups contain many key ingredients that help to improve your health, mainly **vitamins and minerals**.[1] When you are ill, you may not have a good **appetite**. So your **physician** may suggest you drink soup. Both vitamins and minerals are vital to human health and development. They not only prevent certain **diseases**, but also **reverse** them. Some vitamins are needed in order to absorb the minerals. For example, vitamin D is needed to help absorb **calcium**, just as² vitamin C is needed to help absorb **iron**. Vitamin D may be included in soups that contain dairy or fish, and vitamin C can be found in soups that contain tomato or **spinach**.³

A. Fill in the blanks.

| suggest | be vital to | not only...but also... |

1. He _____ that I should finish my work before 11 o'clock.
2. Working hard _____ success.
3. Drinking soup can _____ keep well _____ lose weight.

B. Learn the important grammar and phrase in the text.

1. that 引导的定语从句

- Soups contain many key ingredients that help to improve your health, mainly vitamins and minerals. 汤品中含有很多有助于你健康的重要成分——主要有维生素、矿物质。
- 句中，that help to improve your health 为定语从句，修饰前面的先行词 ingredients，that 是关系代词，在先行词与主语之间起连接作用，并在定语从句中作主语。

练习：The letter that came this morning is from my father.
中文：_____

2. just as 正如

- For example, vitamin D is needed to help absorb calcium, just as vitamin C is needed to help absorb iron. 例如，维生素 D 帮助吸收钙，正如维生素 C 帮助吸收铁。

练习：正如我所料。
英文：_____

3. 被动语态

- Vitamin D may be included in soups that contain dairy or fish, and vitamin C can be found in soups that contain tomato or spinach. 乳制品汤或鱼汤中可能含有维生素 D，维生素 C 可以在含有番茄、菠菜的汤中找到。
- 主语和谓语是被动关系，即主语是动作的承受者时，句中的谓语为被动语态。中文往往用"被""受""给"等被动词来表示被动意义。被动语态结构：be+done。

练习：A grill is the kind of place where grilled food can be ordered.
中文：_____

Task

Group discussion: choose a proper recipe for each person.

1. Joe is short-sighted（近视）. _____ is good for him.
2. Tracy is on diet（规定饮食）. _____ is suitable for her.
3. Susan wants to lower the blood pressure（降低血压）. _____ can help her a lot.
4. Little Tim dreams of growing as tall as Yao Ming. _____ can help him grow up.

pork bone soup	carrot soup	celery soup	tomato soup
calcium	vitamin A	fiber	vitamin C, calorie burning

Further Study

Learn about additional expressions.

English	Chinese	English	Chinese
aioli	蒜泥蛋黄酱	gravy	肉酱
marinade	腌泡汁	soy sauce	酱油
pesto	香蒜酱	custard	牛奶沙司
ketchup	番茄酱	mustard	芥末酱
salad dressing	拌制色拉用的调料	yogurt sauce	酸乳酪
cranberry sauce	蔓越莓沙司	peanut sauce	花生酱
sauce boat	调味汁瓶	chili sauce	辣酱
cream sauce	奶油沙司	blueberry jam	蓝莓酱

Lesson 3

The Kitchen Assistant
厨师助理

Goal

Learn the English names of kitchen equipment and the ways to wash foods in the kitchen.

Warm-up

Write the correct term below each picture.

| unload delivery | insect | detergent |

1. _____ 2. _____ 3. _____

Dialogue

Stella（S）是餐厅的厨师助理，她在和Leo（L）讨论厨师助理的岗位职责。

L: Stella, I heard you landed this job as a kitchen assistant 3 years ago.

S: Yes. The kitchen assistant is a demanding job.

L: What are the duties of kitchen assistant?

S: I help to **unload** deliveries, **unpack** and store food safely.

L: Do you need to clean the kitchen equipment?

S: I clean the kitchen floors and walls, fridges, ovens and work surfaces; wash up or operate the dish-and-glass washing machines; clean pots, pans, electric **cookers**, **scissors** and other equipment by hand; and collect and dispose of wastes.

L: Do you need to do with food?

S: Yes. I have to clean and cut up vegetables, skin and fillet fish, and chop or **mince** meat. And I need to make hot and cold sandwiches, **toast**, soup, desserts, simple salads and fruit dishes, tea and coffee.

L: Well, that's **challenging**.

S: What's more, I need to use equipment such as electric **mixers**, chip machines, knives, cutters, rotate stock and check used-by dates.

A. Match the English with the Chinese.

(　　) 他很幸运，找到了一份好工作。

(　　) 本食品应在其保质期内食用。

(　　) 她很漂亮。除此之外，她心地也很善良。

Unit 1
The Prep Cook

1. He was lucky to land a good job.
2. She is beautiful. What's more, she has a warm heart.
3. Eat the food before the used-by date.

B. Write the correct term below each picture according to the dialogue.

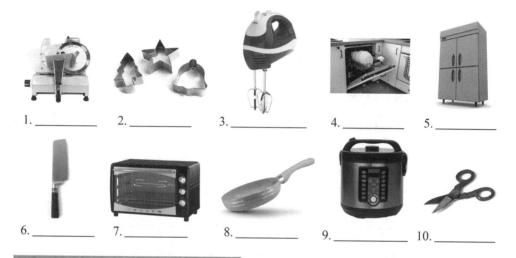

1. _____ 2. _____ 3. _____ 4. _____ 5. _____

6. _____ 7. _____ 8. _____ 9. _____ 10. _____

Role-play

Make a simple dialogue between Stella and Leo about the kitchen assistant's duties.

Reading

If you need to get those fresh fruits and vegetables in your diet off the insects and **chemicals** that come along[1] with them, wash the fruits and vegetables before eating them.[2]

Start with keeping your kitchen **countertops**, refrigerator and cookware clean.

Always wash your hands before preparing meals and **handling** fruits and vegetables.

Keep fresh fruits and vegetables away from **uncooked** meats.

Wait until just before you prepare your fruits and vegetables to wash them. Fruits and vegetables have natural **coatings** that keep **moisture** inside. Washing them will make them **spoil** sooner.³

Gently **rub** fruits and vegetables under running water. Don't use any soaps or chemicals **detergents**.

Firmer fruits and vegetables, such as apples and potatoes, can be **scrubbed** with a vegetable brush while washing with clean water to remove dirt.

A. Reorder the letters.

Letters	Words
o s i m u r e t	m
mi cal che	c
ru s c b	s
s i o p l	s

B. Learn the important grammar and phrase in the text.

1. come along 出现
- If you need to get those fresh fruits and vegetables in your diet off the insects and chemicals that come along with them, wash the fruits and vegetables before eating them. 如果你想食用没有与虫子、化学药品一起出现的水果和蔬菜，要在吃之前把它们清洗干净。

练习：待适当的机会出现，她就能抓住。

提示：right opportunity, take

英文：_____

2. if 引导的条件句
- If you need to get those fresh fruits and vegetables in your diet off the insects and chemicals that come along with them, wash the fruits and vegetables before eating them. 如果你想食用没有与虫子、化学药品一起出现的水果和蔬菜，要在吃之前把它们清洗干净。
- if 引导的条件句表示在某种条件下某事可能发生。

练习：If you have finished eating, you may leave the table.

中文：_____

3. 动名词作主语
- Washing them will make them spoil sooner. 提前清洗容易导致变质。
- 动名词是动词 ing 形式的一种。动名词相当于名词，在句中可作主语、表语、宾语等，也具有动词的特征，可以有自己的状语和宾语。动名词作主语时，谓语动词为单数。

练习：Over-eating often causes indigestion.
中文：_____

Task

Group discussion: what are needed in washing the fruits and vegetables?

Items: _____

a. detergent   b. running water   c. brush   d. tissue   e. soap

Further Study

Learn about additional expressions.

English	Chinese	English	Chinese
kitchen sink	厨房洗涤池	bucket	水桶
faucet	水龙头	grater	磨碎器
hot pot	火锅	masher	捣碎器
toaster	烤面包炉	duster	抹布
air fryer	空气炸锅	cupboard	柜橱

Unit 2

The Pastry Chef

面点组厨师

Lesson 1
The Western Pastry Chef
西式面点师

Goal

Learn to make Western pastries and learn the duties of the pastry chef.

Warm-up

Write the correct term below each picture.

| pastry | mousse cake | pastry bag |
| pastry board | flaky pastry | pastry cutter |

1. _____   2. _____   3. _____

Unit 2
The Pastry Chef

4. _____   5. _____   6. _____

Dialogue

Benton（B）是资深面点师，记者 Ivon（I）在采访他，他们在讨论面点师的职业生涯。

I : Can you tell us your **pastry** chef career steps?

B: I started cooking when I was 8 years old. Back then, I never thought of cooking as a **profession**.

I : Then when did you start to have the idea?

B: When I went to the **culinary** school. After I took my first baking class, I knew it was what I really wanted to do.

I : What advice would you give to those interested in pastry chef jobs?

B: Think hard if you really want to do it, and make sure you have a passion for it.

I : Why do you enjoy the profession?

B: As a pastry chef, I can control my business. I have spare time to try out new recipes.

I : What factors can affect pastry chef's salary?

B: The main factors are skill level, experience and reputation as a cook and baker. Learning cake **decoration** can up pastry chef's salary.

A. Circle Yes or No.

1. As a pastry chef, Benton has no spare time to try out new recipes.   Yes   No
2. Only experience affects the pastry chef.   Yes   No
3. Benton thought of cooking as a profession when he went to the culinary school.   Yes   No

B. Match the English with the Chinese. Draw lines.

1. be interested in...   a. 对……感兴趣
2. have a passion for... b. 业余时间
3. new recipes   c. 对……有激情
4. spare time   d. 新食谱

Role-play

Make a simple dialogue between Benton and Ivon about the pastry chef's career steps.

Reading

Pastry is the name given to various kinds of baked goods made from[1] ingredients such as **flour**, milk, butter, baking **powder** and eggs. Small cakes, pies, **tarts** and other sweet baked goods are called "pastries". Pastry may also refer to the **dough** from which such baked goods are made.[2]

Pastry is different from bread by having a higher fat content. A good pastry is light, but firm enough to support the weight of the **filling**. In other types of pastry, such as Danish pastry and **croissants**, the **characteristic** flaky texture is made by repeatedly rolling out a dough, spreading it with butter, and folding it to produce many thin layers of folds.

A. Reorder the letters.

Letters	Words
f r o u l	f
d e r w o p	p
cha stic rac teri	c
cro ss i ant	c

Unit 2
The Pastry Chef

B. Learn the important grammar and phrase in the text.

1. (be) made from 由……做成
- Pastry is the name given to various kinds of baked goods made from ingredients such as flour, milk, butter, baking powder and eggs. 面点是指用面粉、牛奶、黄油、发酵粉和鸡蛋等原料做成的各种各样的烘烤食品。

练习：葡萄酒是用葡萄酿成的。
英文：_____

2. which 引导的定语从句
- Pastry may also refer to the dough from which such baked goods are made. 面点还可以指代制作这些烘烤食品的生面团。
- 句中，from which such baked goods are made 是定语从句，修饰 dough，which 在定语从句中作介词 from 的宾语，from 来自 be made from。

练习：Read the passage to which I referred in my talk.
中文：_____

Task

Group discussion: the following pictures are some items related to pastry. Guess the Chinese name according to each picture.

1. pastry cream   2. pastry bag   3. pastry cutter   4. pastry board

_____   _____   _____   _____

Further Study

Learn about pastries and pastry tools.

Pastries	Expressions	Pastries	Expressions
	flaky pastry 千层酥饼		pastry fruit pie 水果馅饼
	mousse cake 慕斯蛋糕		Danish pastry 丹麦酥皮甜饼
	French pastry 法式糕点		cookie 曲奇饼干

Pastry Tools	Expressions	Pastry Tools	Expressions
	mixing bowl set 搅拌碗套装		revolving cake tray 旋转蛋糕盘
	whisk 搅拌器		rolling pin 擀面杖
	cake stand 蛋糕架		

Unit 2
The Pastry Chef

Lesson 2
The Chinese Pastry Chef
中式面点师

Goal

Learn the basics of some typical Chinese pastries.

Warm-up

Write the correct term below each picture.

marry girl cake	stinky tofu	cha siu bao
steamed stuffed bun	banana roll	lotus seed cake

1. _____ 2. _____ 3. _____

4. _____

5. _____

6. _____

Dialogue

Tracy（T）是餐厅的中式面点师，她在教 Leo（L）包饺子。

L: Tracy, I want to learn how to make *jiaozi*. Can you give me some tips?

T: My pleasure. At first, you need to get all the ingredients ready. 1 cup of water, 2 teaspoons of salt, 1 pound of lean and ground meat, 2 tablespoons of soy sauce, 1 tablespoon of **sesame** oil, 1 teaspoon of ginger, 1 tablespoon of green onion, 1 teaspoon of black pepper, 1 teaspoon of white pepper and 2 tablespoons of rice wine.

L: Got it. So how can I start?

T: Quite simple. You combine all the ingredients, and then stir.

L: OK. But how can I make the **wrappers**?

T: You can buy wrappers from a supermarket. Take the *jiaozi* wrapper in your hand and put a small spoon of filling in the center of the wrapper. Fold the wrapper in half so it resembles a half circle. **Pinch** the 2 places where the wrapper is folded over. Then, work from one end, and gather a small fold on one side of the wrapper.

L: They are all I need to prepare, right?

T: Not yet. The last step is to cook the *jiaozi* into boiling water in a pot. Stir gently to prevent them from sticking together and then remove the pot from heat.

A. Fill in the blanks.

| wrapper | resemble | prevent...from... |

Unit 2
The Pastry Chef

1. My brother _____ me in looks.
2. We buy _____ and wrap them all up.
3. These rules will _____ accidents _____ happening.

B. The following pictures show the procedures for making mooncakes. Match the picture with the correct procedure.

1. _____ 2. _____ 3. _____ 4. _____

5. _____ 6. _____ 7. _____ 8. _____

9. _____ 10. _____ 11. _____ 12. _____

Procedures:

a. Make the filling.
b. Add the filling.
c. Knead（揉捏）the dough.
d. Break the eggs into another bowl.
e. Press each dough piece into a circle.
f. Brush（刷）each mooncake with the egg wash.
g. Add the sifted（过滤）, dry ingredients to the liquid mixture. Fold in to create a dough.
h. Remove them from the oven.
i. Prepare the mooncake press.

j. Preheat the oven to 374 °F/190 °C. Prepare 2 baking sheets by covering in parchment paper（羊皮纸）.

k. Bake for 30 minutes or until the mooncakes turn golden brown（金褐色）.

l. Make the mooncake dough.

Role-play

Make a simple dialogue between Tracy and Leo about how to make the *jiaozi*.

Reading

Chinese dim sums have a long history and are readily available in gathering places[1], bus stops, **marketplaces** and busy street corners in China and in places such as New York City's Chinatown. The followings show some **typical** Chinese dim sums.

Steamed Buns

Steamed buns, or *baozi*[2], are a kind of **staple** food in China for breakfast or any other time in a day. People usually come with **savory** filling that is added to the dough before steaming.

Chicken Feet

Because they are mostly skin and **tendons**, the snack has a **chewy** texture with small bones.[3] They usually are marinated in a **spicy** chili sauce.

Stinky Tofu

Stinky tofu, also known as *choudoufu*, is a Chinese snack famous for[4] its **horrible** aroma. Stinky tofu is served as street food.

A. Read the text again and circle Yes or No.

1. Chinese dim sums don't have a long history.   Yes   No
2. Steamed buns, or stinky tofu, is a kind of staple food in China.   Yes   No
3. Chicken feet are marinated in a spicy chili sauce.   Yes   No

Unit 2
The Pastry Chef

B. Learn the important grammar and phrases in the text.

1. gathering place 聚集场所
- Chinese dim sums have a long history and are readily available in gathering places, bus stops, marketplaces and busy street corners in China and in places such as New York City's Chinatown. 中式小吃有着悠久的历史，在中国的聚集场所、车站、市场、繁华街头以及如纽约市中国城等地方都可以方便地买到。
- 动名词作定语表示用途，如：dining room 餐厅。

练习：十二月三十一日，人们最喜欢的聚集场所是时代广场。

提示：Times Square

英文：_____

2. 音译
- steamed buns 与 *baozi* 均指蒸制的面食，但 *baozi* 特指中国的传统包子，其内馅丰富多样，如猪肉大葱、三鲜等，体现中式面点的精致与讲究。Stinky tofu 与 *choudoufu* 均指臭豆腐，但 *choudoufu* 由中文音译而来。前文提到的 dumpling，描述的是一种国内外普遍存在的由面皮包裹馅料的食物，但 *jiaozi* 则特指中国的传统饺子，承载着团圆、辞旧迎新的文化意义。随着中华优秀传统文化的传播，越来越多的外国友人喜爱中华传统美食，习惯使用由中文音译的食物名称。

练习：我们在端午节吃粽子。

提示：Dragon Boat Festival

英文：_____

3. because 引导的原因状语从句
- Because they are mostly skin and tendons, the snack has a chewy texture with small bones. 凤爪基本上是由皮和筋构成的，所以骨小、质地耐嚼。
- because 译为"因为"，Because they are mostly skin and tendons 为原因状语从句，the snack has a chewy texture with small bones 为主句。because 引导的从句置于主句前表示强调。

练习：I did it because he told me to.

中文：_____

4. (be) famous for 以……而著名
- Stinky tofu, also known as *choudoufu*, is a Chinese snack famous for its horrible aroma. 臭豆腐是以味臭而著名的中式小吃。

练习：这个地区以"康沃尔郡度假胜地"而著名。

提示：Cornish Riviera

英文：_____

Task

Choose the corresponding picture for each item mentioned in the text.

1. steamed buns _____
2. chicken feet _____
3. stinky tofu _____

a.   b.   c.   d.   e.

Further Study

A. Learn about types of pastry.

wife cake　　　　　pineapple pastries　　　　leek pie
老婆饼　　　　　　凤梨酥　　　　　　　　　韭菜饼

almond biscuits　　　　　steamed dumplings
杏仁饼　　　　　　　　　蒸饺

B. Learn about Chinese dim sums.

shrimp dumplings　　　potstickers　　　　*zongzi*
虾饺　　　　　　　　锅贴　　　　　　　粽子

Unit 2
The Pastry Chef

preserved egg and
pork porridge
皮蛋瘦肉粥

Shanghai
steamed buns
上海小笼包

Unit 3

The Cold Dishes Chef

冷菜组厨师

Lesson 1
The Chef Garde Manger
冷房厨师

Goal

Learn to make cold foods.

Warm-up

Write the correct term below each picture.

| lettuce   coconut   mango   pineapple   eggplant   kiwi fruit |

1. _____　　2. _____　　3. _____

4. _____　　5. _____　　6. _____

Unit 3
The Cold Dishes Chef

Dialogue

Wendy（W）是餐厅的冷房厨师，她在教 Leo（L）做冷汤（黄瓜汤）。

L : Wendy, can you teach me how to make cold soup?

W: OK. How about **cucumber** soup?

L : Sounds good.

W: You need 1 cucumber, 14 fresh **mint** leaves and 2 cups of **buttermilk**.

L : How do I deal with the cucumber?

W: Just chop it into pieces. You also need 2 tablespoons of **yogurt**, 2 tablespoons of olive oil, salt and pepper.

L : And then what can I do?

W: Reserve 4 mint leaves for the presentation. Chop the rest of the mint and place along with the chopped cucumber in a food processor. Blend until smooth.

L : Got it. And I will continue processing and add the buttermilk, yogurt and olive oil to blend into a smooth liquid.

W: Don't forget to add the pepper and salt.

L : And the last step is to decorate with the 4 mint leaves.

W: That's right.

A. Match the English with the Chinese. Draw lines.

1. Yogurt is very nutritious.　　　　　　a. 炼乳很甜。
2. Chop the carrots up into small pieces.　b. 酸奶很有营养。
3. Condensed milk is very sweet.　　　　c. 把胡萝卜切成小块。

B. List the ingredients and utensils needed for making the cucumber soup according to the dialogue.

Ingredients: _____

a. pumpkin   b. banana   c. buttermilk   d. cucumber   e. pepper

f. mint leaf   g. olive oil   h. salt   i. strawberry   j. yogurt

Utensils: _____

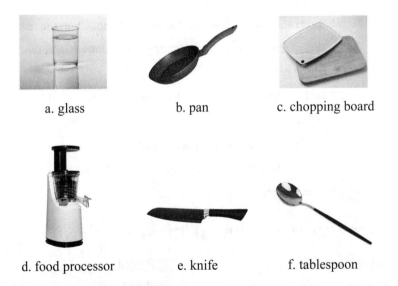

a. glass   b. pan   c. chopping board

d. food processor   e. knife   f. tablespoon

Role-play

Make a simple dialogue between Wendy and Leo about how to make the cold soup.

Unit 3
The Cold Dishes Chef

Reading

It's easy to[1] create a professional looking of a fruit **platter**. The key to doing[2] this is to cut and slice your fruit into **similar** size and **shape** pieces. For this platter, choose a **melon**, strawberries and **mangoes**.

To create a heart shape in the center of this fruit platter, cut off the green leaves from the strawberries, slice them up to the top (without cutting through the end), and **fan** out the berries.

Begin to **assemble** the platter by shaping the sliced strawberries into a heart shape.

Continue to slice other fruit. With mangoes, peel the fruit first, then slice down each side to remove the **flesh** from the center **pit**. Slice into **equal** size pieces. Cut melons in half, remove seeds, and slice off rind. **Proceed** to make equal size slices of fruit.

Arrange fruit around the center heart. Keep slices of each fruit variety together, and fans out on the platter.

A. Fill in the blanks.

proceed to	in half	begin to	slice down

1. I _____ take the course of culinary art this term.
2. She _____ a piece of meat.
3. He cut the watermelon _____.
4. Let's _____ next topic.

B. Learn the important phrases in the text.

1. (be) easy to do sth. 做某事比较容易
- It's easy to create a professional looking of a fruit platter. 看上去很专业的水果拼盘比较容易做。

练习：那个地方比较容易到达。

英文：_____

2. key to doing sth. 做某事的关键
- The key to doing this is to cut and slice your fruit into similar size and shape pieces. 关键是将水果大小、形状切得类似。
- 这句话中的第一个 to 是介词，第二个 to 是不定式。

练习：经济增长是解决贫困问题的关键。

提示：economic growth, poverty

英文：_____

Task

Group discussion: learn to describe the fruit carving art. Keep the steps in a right order according to the following pictures.

Right order: _____

1. Trim（抠去）away the core（苹果核），leaving the flesh and skin.
2. Cut notches（切口）along the edges of the leaf.
3. With the carving knife, cut the slice（切下一片）to the shape of a leaf.
4. Wash the apples.
5. Cut a wedged-shape（楔形）slice as in the picture.
6. With the tip of the knife, make curving grooves（槽）in the skin to represent the veins（叶脉）of the leaf. Work from the base of the leaf to the tip.

Unit 3
The Cold Dishes Chef

Further Study

Learn about carving fruits and their names.

apple swan   watermelon peacock   watermelon basket   flower on apple

Lesson 2
The Glacier Chef
冷冻菜厨师

Goal

Learn to make cold desserts and decorate dishes.

Warm-up

Write the correct term below each picture.

| whisk | fence | candy | zest | fireplace | chimney |

1. _____ 2. _____ 3. _____

4. _____ 5. _____ 6. _____

Unit 3
The Cold Dishes Chef

Dialogue

May（M）是餐厅的冷冻菜厨师，她在教 Leo（L）做柠檬酸奶奶油。

L : May, can you teach me how to make the lemon yogurt cream?

M: Yes. You have to prepare plain creamy yogurt, 4 eggs, 100g butter, 2 lemons, 110g sugar, 1 tablespoon of honey, 4 slices of **gingerbread**, 2 pinches of **nutmeg**.

L : OK. Then how can I start?

M: Wash the lemons. Remove the **zests** and cut it in **strips**. Place them in cold water. Bring to a boil for 3 minutes. **Rinse** under running water and drain them. Heat 2 spoons of water along with 30g sugar. Meanwhile, add the zests in candy over low heat for 8 to 10 minutes. Set aside on a plate.

L : Got it. Is the next step to cut 4 disks out of the gingerbread and place the disks in the bottom?

M: Yes, beat the yogurt with honey and 2 pinches of nutmeg and divide in the **ramekins** and store in the fridge for 1 hour. Can you tell me what would you do next?

L : Press the lemons.

M: Yes. In a bowl, beat the juice along with eggs and the remaining sugar, and place the bowl in the bain-marie and beat by an electric **whisk** until the mixture is firm and foamy. Get it out of the bain-marie. Add the butter in cubes. And then cool it a bit.

L : I know the last step is to pour this lemony cream over yogurt. Smooth the surface then leave it for 3 to 4 hours in the fridge.

A. Reorder the letters.

Letters	Words
ss t er de	d
m o e n l	l
l i a n p	p
tu mix re	m

B. Choose the ingredients for the lemon yogurt cream according to the dialogue.

Ingredients: _____

a. blue cheese   b. butter   c. chocolate   d. egg   e. gingerbread

f. honey   g. lemon   h. nutmeg   i. sugar   j. yogurt

Role-play

Make a simple dialogue between May and Leo about how to make the lemon yogurt cream.

Reading

It's a wonderful idea to make some **creative** desserts as gifts for your friends. The followings are some **tips** on how to decorate[1] a plain gingerbread into a cute gingerbread house.

Unit 3
The Cold Dishes Chef

（**icicle**）Use a round decorating tip. With heavy to light **pressure**, **pipe** from top to **bottom**.

（fence）Place **pretzel** sticks side by side. Use icing in bag fitted with a round decorating tip to secure rails. Let them completely set.

（fireplace/chimney）Outline fireplace/chimney area with a round decorating tip.

（window）Outline window with a round decorating tip. Attach sugar **wafer** cookies to window[2] with icing.

A. Translate the following sentences into Chinese.

1. What's the function of the round decorating tip?
   _____

2. They examine the place from top to bottom.
   _____

3. The two desks stand side by side.
   _____

B. Learn the important phrases in the text.

1. how to do sth. 做某事
   - The followings are some tips on how to decorate a plain gingerbread into a cute gingerbread house. 下面是一些关于如何把一块普通的姜饼装饰成可爱姜饼屋的技巧。

   练习：我教你如何煮方便面。
   提示：instant noodles
   英文：_____

2. attach...to... 将……粘在……上
   - Attach sugar wafer cookies to window with icing. 用糖衣将威化饼干粘在窗户上。

   练习：他会将标签粘在你的行李上。
   提示：label, luggage
   英文：_____

Task

Group discussion: what are they?

| icing | waxed paper | round decorating tip |

 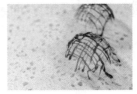 

a. _____   b. _____   c. _____

Further Study

Learn about the categories of cold desserts.

Generally speaking, there are 3 categories of cold desserts: puddings, cake-based desserts, and pies. The following pictures show the common desserts of 3 categories.

- Puddings

tapioca pudding　　　　banana pudding　　　　rice pudding
西米布丁　　　　　　　香蕉布丁　　　　　　　大米布丁

- Cake-based desserts

trifle　　　　　　　　flan　　　　　　　strawberry shortcake
乳脂松糕　　　　　　果馅饼　　　　　　草莓酥饼

Unit 3
The Cold Dishes Chef

- Pies

orange-cream pie
橙子奶油馅饼

fruit-topped tart
水果蛋挞

lemon curd tart
柠檬酱蛋挞

Unit 4

The Hot Dishes Chef

热菜组厨师

Lesson 1
The Fry Chef
煎炸厨师

Goal

Learn the basic recipes of Chinese and Western fried foods.

Warm-up

Write the correct term below each picture.

| green bean | ginger | mushroom | shrimp | turnip | pea |

1. _____ 2. _____ 3. _____

4. _____ 5. _____ 6. _____

Unit 4
The Hot Dishes Chef

Dialogue

Helen（H）是餐厅的煎炸厨师，她在教 Leo（L）做蔬菜蛋糕。

L: Helen, can you do me a favor?

H: Well, what do you want me to do?

L: I want to create a dish for vegetarians. Can you give me some suggestions?

H: How about vegetable cakes?

L: Sounds interesting. What are the ingredients?

H: Some carrots, **turnips**, peas and green beans.

L: These vegetables are nutritious.

H: Right. And cut the vegetables after draining.

L: OK.

H: Pat dry by wrapping the vegetables in a dry cloth.

L: It's easy.

H: **Flour** the vegetables.

L: And bake them?

H: No, far away from that step. Take a bowl, mix the clarified butter with the eggs. Add the flour, baking powder, **citrus** zests, **cinnamon** and seasons.

L: Oh, my mistake. And now it is time to bake them, right?

H: Yes. Bake at 250 ℃ for 10 minutes, then lower the temperature to 170 ℃ and bake for another 45 minutes. When it is finished, **unmold**, slice and eat it warm.

A. Reorder the letters.

Letters	Words
ve ta ge rian	v
na mon cin	c
son sea	s
l i e c s	s

B. The followings are methods of Chinese cooking. Match the term with the corresponding picture.

| stir-fry | roast | boil | stew | fry |

1. _____
2. _____
3. _____
4. _____
5. _____

Role-play

Make a simple dialogue between Helen and Leo about how to make the vegetable cakes.

Reading

A steak is a cut of meat (usually beef).

The cooking time of a steak is based upon personal **preference**; shorter cooking time retains more juice, while longer cooking time results in[1] drier and **tougher** meat. The following **terms** are in order from least cooked to most cooked:

1. **Raw** — Uncooked.
2. Blue rare or very rare — Cooked very quickly. The steak will be red in the inside

Unit 4
The Hot Dishes Chef

and barely warmed.

3. **Rare** — (52 ℃) The outside is grey-brown, and the middle of the steak is red and slightly warm.

4. **Medium rare** — (55 ℃) The steak will be fully red, of which center is warm.[2] This is the standard degree of cooking at most steakhouses.

5. **Medium** — (60 ℃) The middle of the steak is hot and red with pink center. The outside is grey-brown.

6. **Medium well done** — (65 ℃) The meat is light pink surrounding the center.

7. **Well done** — (71 ℃ above core temperature) The meat is grey-brown **throughout** and slightly **charred**.

8. **Overcook** — (much more than 71 ℃) The meat is dark throughout and slightly bitter.

A. Fill in the blanks.

| a cut of | base upon | grey-brown |

1. This is a story _____ historical factor.
2. Would you like _____ pork?
3. The color of her hair is _____.

B. Learn the important grammar and phrase in the text.

1. result in 导致，则

- ...while longer cooking time results in drier and tougher meat. ……时长则肉干而老。

练习：我们努力，则此事成功。

英文：_____

2. 介词+which 引导的定语从句

- The steak will be fully red, of which center is warm. 牛排呈血红色，里面也煎得温热。

- 关系代词 which 在定语从句中用作介词宾语时，介词既可置于从句之首，亦可置于从句之末。但以置于从句之首较为正式。"介词+which"结构中，which 不可省略。

练习：I have three cooking books, the first of which I like best.

中文：_____

Task

Write down the English terms according to the Chinese meanings.

1. 全生：_____
2. 近生：_____
3. 三分熟：_____
4. 煎焦：_____
5. 七分熟：_____
6. 一分熟：_____
7. 全熟：_____
8. 五分熟：_____

Further Study

Learn about the types of beef.

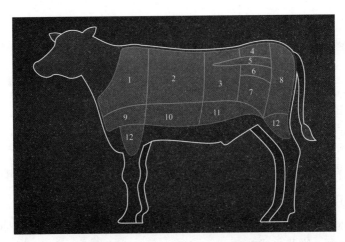

1. chuck 牛肩胛肉
2. rib 牛肋骨
3. short loin 牛前腰脊肉
4. sirloin 牛后腰脊肉
5. tenderloin 牛里脊肉
6. top sirloin 牛上后腰脊肉
7. bottom sirloin 牛下后腰脊肉
8. round 牛腹腿
9. brisket 牛胸肉
10. plate 腹肉
11. flank 侧腹
12. shank 牛腱

Lesson 2
The Saute Chef
烹炒厨师

Goal

Learn to make saute food.

Warm-up

Write the correct term below each picture.

| pasta fork | saute tongs | frying pan |
| meat lifter | saute pan | lid |

1. _____

2. _____

3. _____

4. _____   5. _____   6. _____

Dialogue

Leo（L）进入新部门学习。餐厅的烹炒厨师 Anna（A）让 Leo 去买个合适的锅，并教他做法式炒菜。（shop assistant=S）

S: Hi, I'm Dick Fisher. Welcome to our kitchenware store.

L: I'm looking for pans.

S: What kind of pans do you want, **saute** pans or frying pans?

L: What's the difference?

S: They both have **flat** bottoms, but a saute pan has a lid and vertical sides so that the food can be shaken in the pan without spilling.

L: I need a saute pan.

A: Leo, today I'll show you how to make French style saute food.

L: What's that?

A: You take whatever meat item and season it. **Dredge** it in flour. Then **dip** in a mixture of beaten eggs, chopped **parsley** and **grated** cheese.

L: And then?

A: Then you saute the item in butter.

L: That's new to me. How does it taste?

A: It tastes wonderful. The food is so popular among young people in France!

Unit 4
The Hot Dishes Chef

A. Choose the ingredients needed in making French style saute food mentioned in the dialogue.

Ingredients: _____

a. cheese   b. egg   c. flour   d. beef   e. parsley

B. Match the English with the Chinese. Draw lines.

a. 她在找她的小孩。

b. 家附近有家厨具店很方便。

c. 中国菜味道好极了。

d. 这是纯羊毛的，因此手感柔和。

1. It's very convenient to have a kitchenware store near the house.

2. She is looking for her child.

3. It's made of pure wool, so it's very soft.

4. Chinese food tastes wonderful.

Role-play

Make a simple dialogue between Anna and Leo about how to make the French style saute food.

Reading

Some people say that this dish is the best dish of Shanghai. Its Chinese name means **crystal shrimps**. Although it looks like a dish simply made of shrimps, the **material**'s **processing** and cooking can be quite **complicated**.[1] What makes the dish good is that each **bite** first gives **crisp** feeling at the skin of the shrimps, but when you bite **further** you will soon feel the tender and **juicy** from the shrimps' body. To make the shrimps taste like this depends on the skills of the cook and how much time the cook spends.[2] The dish can be found in a lot of Shanghai restaurants.

A. Fill in the blanks and match the word with the correct definition.

tender	juicy	processing

1. Easily chewed or cut. _____
2. A series of operations performed in the making or treatment of a product. _____
3. Full of juice. _____

B. Learn the important grammar in the text.

1. although 引导的让步状语从句
- Although it looks like a dish simply made of shrimps, the material's processing and cooking can be quite complicated. 虽然这道菜看起来是用虾仁简易制成的，但食材处理和烹制的过程相当复杂。
- although 译为"虽然"，与 though 可互换使用，二者都能与 yet、still 连用，但不能与 but 连用。句中，Although it looks like a dish simply made of shrimps 为 although 引导的让步状语从句，the material's processing and cooking can be quite complicated 为主句。

练习：Although my uncle is old, he looks very strong and healthy.

中文：_____

2. 不定式（to do）作主语
- To make the shrimps taste like this depends on the skills of the cook and how much time the cook spends. 要达到这般效果就要看厨师在烹饪时对火候和时间的掌控技巧。
- 不定式（to do）作主语常表示某次具体的行为，相当于名词或代词。谓语动词通常为单数形式。

练习：To get there by bike will take us half an hour.

中文：_____

Unit 4
The Hot Dishes Chef

Task

Group discussion: the following pictures are some famous saute dishes in Chinese food. Match the picture with the English name.

| sauteed wild mushrooms | sauteed ox liver | sauteed sweet potatoes |
| sauteed green beans | sauteed broccoli with garlic | |

a. 炒红薯　　　b. 炒野菌菇　　　c. 炒豆角　　　d. 炒牛肝　　　e. 蒜蓉西蓝花

Further Study

Learn to be a qualified saute chef.

The saute chef is responsible for（负责）sauteing all kinds of vegetables, meat and seafood. To make saute food successfully, the saute chef first should chop the ingredients（切菜）, and then heat the pan（热油锅）. Add some oil. After waiting for a minute, stir regularly（不时）the food in the pan. And finally take off the heat and pour onto some kitchen towel to dab off the excess oil（多余的油）.

Lesson 3

The Grill and Roast Chef
烧烤厨师

Goal

Learn the expressions about grilling and roasting foods.

Warm-up

Write the correct term below each picture.

| pancake | spring onion | roast duck |
| baking pan | meat thermometer | chopping board |

1. _____
2. _____
3. _____
4. _____
5. _____
6. _____

Unit 4
The Hot Dishes Chef

Dialogue

Dora（D）是餐厅的烧烤厨师，她在教 Leo（L）做烤鸭。

L: Dora, can you teach me how to make roast duck?

D: If you want to make roast duck, you have to prepare duck breasts, garlic cloves, salt and pepper.

L: OK.

D: Slash the fat side in parallel lines. Add salt and pepper on both sides.

L: I know. This step aims to add flavor to the duck.

D: That's right. And you should slide in a bit of fresh garlic between fat and meat, then **lace** them meat to meat and tie them together.

L: OK. Where can I put them?

D: Place them in a baking pan.

L: How can I bake them?

D: Bake them 15 minutes at 250 ℃. In the 7th minute after you begin baking, you should **degrease** the pan and add a glass of water.

L: That is to say, I degrease the pan and add water halfway.

D: Yes. When it is finished, remove the strings from the breasts, and slice them finely. The **doneness** must be rare.

L: I will have a try this evening.

A. Circle Yes or No.

1. Dora is a saute chef.   Yes   No
2. Dora is teaching Leo how to make roast goose.   Yes   No
3. Leo will practice this dish tomorrow.   Yes   No

B. Choose the ingredients for the roast duck according to the dialogue.

Ingredients: _____

a. chicken   b. duck   c. garlic clove   d. goose   e. lettuce
f. pepper   g. spring onion   h. salt   i. brown sugar   j. a glass of water

Role-play

Make a simple dialogue between Dora and Leo about how to make the roast duck.

Reading

Peking Duck, or Peking **Roast** Duck is a **famous** dish in Beijing that has been prepared since the **imperial era**, and now is considered as one of Chinese national foods.[1]

The dish is **prized** for the thin, **crispy** duck skin, sliced in front of the diners by the cook. Ducks are seasoned before being roasted in a closed or hung oven.[2] The meat is eaten with pancakes, spring onions, and hoisin sauce or sweet bean sauce. The 2 most notable restaurants in Beijing which serve this delicacy are *Quanjude* and *Bianyifang*, 2 centuries old[3] establishments which have become **household** names.[4]

A. Translate the following sentences into Chinese.

1. After watching the Peking Opera, we can have some Peking Roast Duck.

2. He is considered as a good boy.

3. He is prized for his wonderful performance on the stage.

Unit 4

The Hot Dishes Chef

B. Learn the important grammar and phrase in the text.

1. 完成时+since
- Peking Duck, or Peking Roast Duck is a famous dish in Beijing that has been prepared since the imperial era, and now is considered as one of Chinese national foods. 北京烤鸭是北京的一道名菜，始于帝国时代，现已被认为是中国的国菜之一。
- since 译为"自从"，句中谓语动词通常用完成时，表示动作或状态从过去某一时刻开始，一直持续到现在，还可能持续下去。

练习：I have lived here since January.
中文：_____

2. before being done
- Ducks are seasoned before being roasted in a closed or hung oven. 鸭子在放入焖炉或挂炉烘烤之前涂上调料。
- before 是介词，其后可跟动名词。be done 为被动语态，放在 before 后，变为 being done。

练习：In 1906, laws were passed to make sure that meat was inspected before being sold.
中文：_____

3. centuries old 数百年的
- The 2 most notable restaurants in Beijing which serve this delicacy are *Quanjude* and *Bianyifang*, 2 centuries old establishments which have become household names. 北京最著名的供应烤鸭的两家店便是全聚德和便宜坊，这两家店都已有两百年的历史，现已成为家喻户晓的名字。

练习：百年老店，久享盛名。
英文：_____

4. 烤鸭技艺
- 全聚德的挂炉烤鸭技艺和便宜坊的焖炉烤鸭技艺，作为中国传统烹饪艺术的瑰宝，于 2008 年被列入第二批国家级非物质文化遗产名录。这两项技艺不仅代表了北京烤鸭的独特风味，更承载了深厚的文化底蕴和历史传承。它们以其独特的制作工艺、精湛的烹饪手法和无与伦比的美味，赢得了国内外食客的高度赞誉。国家级非物质文化遗产名录涵盖了众多传统技艺和民俗文化，展现了中华民族丰富多彩的文化遗产。

Task

Read the text again and circle Yes or No.

1. Peking Roast Duck is a famous dish in Beijing. Yes   No
2. Ducks are roasted in closed ovens only. Yes   No
3. *Quanjude* and *Bianyifang* have become household names. Yes   No

Further Study

Learn about grill utensils.

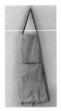 apron 围裙

 bacon grill press 培根烧烤压板

 mitt 烤箱手套

 brush 刷子

 fork 叉子

 grill rack 烧烤架

 grill smoker box 烧烤烟熏炉

 spatula 铲子

 tongs 钳子

 tool holder 置物架

Unit 4
The Hot Dishes Chef

Lesson 4
The Vegetable Chef
蔬菜厨师

Goal

Learn the English names of various kinds of vegetables and their nutritious ingredients.

Warm-up

Write the correct term below each picture.

| broccoli | sweet potato | cabbage |
| lotus root | aloe | spinach |

1. _____
2. _____
3. _____
4. _____
5. _____
6. _____

Dialogue

Susan（S）是餐厅的蔬菜厨师，她在教 Leo（L）做美式番茄碎。

L: Susan, can you teach me how to make **crushed** tomato?

S: Crushed tomato is a kind of vegetable-based sauce. And it's very nutritious. The more tomatoes one eats, the lower the risk of **cancer**.

L: What a good vegetable!

S: If you want to make crushed tomato, you should prepare fresh tomatoes, onions, garlic cloves, olive oil, salt, pepper and sugar.

L: OK. I know the first thing to do is to **peel** the onions and slice them finely.

S: Yes. And in a large pot, heat up the olive oil. Add the onions and let it sweat without browning for a few minutes.

L: OK. Then what can I do?

S: Add the peeled, seeded and crushed tomatoes. Season them with salt, pepper, garlic and a bit of sugar.

L: OK. Is that all?

S: Cook over medium heat for about 30 minutes. The cooking time depends on the tomatoes. Their water must have **evaporated** at the end.

A. Match the English with the Chinese. Draw lines.

1. I don't want to take the risk of losing it.
2. Our plans depend on the weather.
3. She seasoned the fish with sugar.

a. 她用糖给鱼调味。
b. 我不想冒失去它的风险。
c. 我们的计划取决于天气。

Unit 4
The Hot Dishes Chef

B. Choose the ingredients for the crushed tomato according to the dialogue.

Ingredients: _____

a. carrot   b. garlic   c. olive oil   d. pepper   e. potato

f. pumpkin   g. onion   h. sweet potato   i. tomato   j. orange

Role-play

Make a simple dialogue between Susan and Leo about how to make the crushed tomato.

Reading

Among many types of foods, **beans** are perhaps almost the perfect type of food. Beans, in fact, are so filled with **nutrition** and so reduced in **calories** that, if you happen to be **dieting**, beans are the right food for you.[1] Beans are one of the best **sources** of **fiber**. Fiber has been shown to help us prevent many illnesses, including heart disease and cancer.[2]

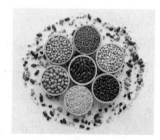

One of the most well-known beans is the black bean. It is also one you can use in several types of recipes. The black bean is a source of **dietary** fiber. This fiber contained in black beans and other beans has the **effect** of keeping blood sugar level down[3] after a big meal.

A. Fill in the blanks and make changes where necessary.

| happen to be | be filled with | a source of |

1. The fridge _____ foods.
2. Fish is considered as _____ protein.
3. If you _____ in town, would you ask about it?

B. Learn the important grammar and phrase in the text.

1. so...that... 引导的结果状语从句

- Beans, in fact, are so filled with nutrition and so reduced in calories that, if you happen to be dieting, beans are the right food for you. 事实上，豆类不但营养丰富，而且低卡路里。若你正巧在节食，豆类正是适合你的食品。
- so...that... 表示"如此……以至于……"。so 后一般跟形容词或副词。...that, if you happen to be dieting, beans are the right food for you 为表示结果的状语从句。

练习：He was so ill that we had to send for doctor.

中文：_____

2. 现在完成时

- Fiber has been shown to help us prevent many illnesses, including heart disease and cancer. 研究表明，纤维素有助于我们预防包括心脏病和癌症在内的许多疾病。
- 现在完成时表示过去发生的动作对现在造成的影响或结果。谓语动词形式为 have/has done。

练习：I've finished my homework.

中文：_____

3. keep sth. down 控制

- This fiber contained in black beans and other beans has the effect of keeping blood sugar level down after a big meal. 黑豆和其他豆类中的膳食纤维有控制大餐后体内血糖水平的效果。

练习：这些药片是控制发热的。

提示：tablet, fever

英文：_____

Unit 4
The Hot Dishes Chef

Task

Group discussion: according to the following charts, put the vegetables in order from the most to the least based on the instructions.

Picture	Item	Serving	Fat	Fiber	Protein	Carbon
	beans	1 cup	17 g	11 g	33 g	28 g
	broccoli	1 bunch	2 g	18 g	18 g	32 g
	cabbage	1 kilogram	30 g	10 g	15 g	46 g
	mushrooms	1 cup sliced	0 g	1 g	1.5 g	3.2 g
	peas	1 cup	0.5 g	7 g	8 g	21 g
	spinach	1 bunch	1 g	9 g	9.5 g	12 g
	tomato	1 kilogram	2 g	10 g	9 g	35 g

1. Fat: _____
2. Fiber: _____
3. Protein: _____
4. Carbon: _____

Further Study

Learn about additional expressions.

English	Chinese	English	Chinese
Chinese cabbage	大白菜	asparagus	芦笋
lettuce	生菜	cucumber	黄瓜
celery	芹菜	sponge gourd	丝瓜
leek	韭菜	pumpkin	南瓜
cauliflower	菜花	white gourd	冬瓜
greens hearts	菜心	taro	芋头
kale	羽衣甘蓝	eggplant	茄子
sprouts	豆苗	radish	小萝卜
bamboo shoot	笋	carrot	胡萝卜

Unit 5

The Larder Chef

肉类组厨师

Lesson 1

The Butcher
肉类处理工

Goal

Learn the duties of the butcher and meat cutter, and learn to process meat.

Warm-up

Write the correct term below each picture.

burger press	butcher	block brush
sausage	double-edged cleaver	pork

1. _____

2. _____

3. _____

Unit 5
The Larder Chef

4. _____   5. _____   6. _____

Dialogue

为了更好地熟悉肉食品加工部门的工作内容，Leo（L）向 Jason（J）咨询其所做的工作。

L: What does a butcher do, Jason?

J: In general, butchers are responsible for **boning** and cutting up meat.

L: How did you become a butcher?

J: I'm more like self-taught. I started laboring at the **abattoirs** and went from there... I just had been boning out beef for 20 years, and then I started doing a little part-time job in butcher shops and I went along learning a bit.

L: Is it dangerous?

J: It can be, yeah, when you're working with band **saws**, mincers... You've got to treat the machinery with the greatest respect because once you slip, you could just cut your hand off.

L: Are there any tips for getting a job as a butcher?

J: If you take on an **apprenticeship**, don't expect to leap into shop-keeping right away. Like everything, you start at the bottom, sweeping the floors, cleaning the blocks, and you just generally learn as you go along. But after you've done your apprenticeship, you're ready to go.

A. Reorder the letters.

Letters	Words
u t h b c e r	b
pp a tice ren ship	a
s re ct pe	r
nery ma chi	m

B. Name the following meat.

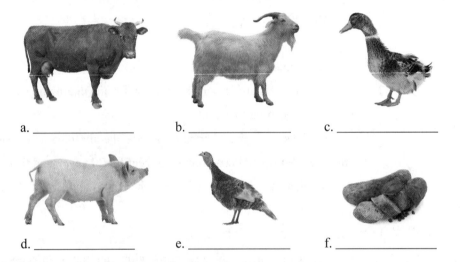

a. _____ b. _____ c. _____

d. _____ e. _____ f. _____

Role-play

Make a simple dialogue between Jason and Leo about the butchers' duties.

Reading

There are 2 **cuts** of meat: tough and **tender**. Tough cuts of meat contain the **muscle**, which requires braising or stewing to become tender.[1] Tender cuts demand quick cooking to retain their **texture** and seal in their flavor.

Unit 5

The Larder Chef

Select high-quality[2] beef

Choose beef with minimal **outer** fat. The meat should be firm, fine-textured and a light cherry red.

Choose high-quality pork

Select pork that's **pinkish**-white to pink in color and firm to the touch.

Choose high-quality lamb

Look for[3] **lamb** that's been butchered at 5 to 7 months or younger. It has a more delicate flavor and texture than mutton, which takes on a rich **gamy** flavor.

A. Match the English with the Chinese. Draw lines.

1. The old factory has taken on a new look.　　a. 这孩子的脸蛋红扑扑的。
2. Fry quickly to seal in the flavor of the meat.　　b. 这家老厂已呈现出一派新面貌。
3. The child has cherry red cheeks.　　c. 快速把肉用油煎一下,以留住其美味。

B. Learn the important grammar, word, and phrase in the text.

1. which 引导的非限定性定语从句
- Tough cuts of meat contain the muscle, which requires braising or stewing to become tender. 老的肉包含肌肉，需要炖或煮来使其变嫩。
- 非限定性定语从句起补充说明作用，在句中并非必不可少，即使省略也不会引起意义上的混乱，它与主句往往用逗号隔开。

练习：It offers food of Hunan flavor, which is hot and spicy.

中文：_____

2. high-quality 优质的
- Select high-quality beef 选择优质牛肉

练习：优质矿泉水已通过国家级鉴定。

提示：state-level

英文：_____

3. look for 寻找
- Look for lamb that's been butchered at 5 to 7 months or younger. 选择屠宰时为5~7个月或更小的羊的羔羊肉。

练习：你仍在寻找工作吗?

英文：_____

Task

Read the text again and circle Yes or No.

1. Tender meat requires long cooking such as braising and stewing.   Yes   No
2. The dark cherry red beef should be selected.   Yes   No
3. The younger lamb has more delicate texture than the older one.   Yes   No

Further Study

Learn about additional expressions.

English	Chinese	English	Chinese
poultry	禽肉	butcher shops	肉店
lamb	羔羊肉	fish markets	鱼市
bacon	培根	primal cuts	大块肉
sausage	香肠	trimmings	肉丝
dried floss	肉松	ground meat	肉末
ham	火腿	slaughter	屠宰

Lesson 2

The Fish and Seafood Chef
海鲜厨师

Goal

Learn the terms about fish and seafood.

Warm-up

Write the correct term below each picture.

| lobster | crab | shrimp | salmon | oyster | octopus |

1. _____
2. _____
3. _____
4. _____
5. _____
6. _____

Dialogue

Aida（A）是餐厅的海鲜厨师，她在教 Leo（L）制作荷兰酸酱拌水煮三文鱼。

L: The **poached salmon** with hollandaise sauce tastes wonderful. Aida, how do you make it?

A: Thank you. It's not complicated. First, you should prepare 3 tablespoons of fresh lemon juice, 1 tablespoon of olive oil, 1 cup of butter, 2 skinless, boneless salmon **fillets**, 3 egg yolks, and salt and pepper.

L: OK. Then what to do next?

A: Find a pan and pour lemon juice and olive oil into it, as well as enough water. Season the water to taste with salt and pepper, then add the salmon.

L: When do you put the salmon?

A: Place the salmon over medium-high heat, and heat until the water is hot and steaming.

L: What's the function of egg yolks?

A: Place the egg yolks in a metal bowl and whisk in hot water. Place the bowl over, but not touch the boiling water. Whisk constantly until the yolks **thicken**.

L: When to put the butter?

A: When the yolks have thickened, begin whisking in the butter, a cube at a time, until it melts and **incorporates** into the hollandaise sauce. Then remove the heat, whisk in the lemon juice, and season to taste with salt and pepper.

L: Sounds easy.

A: To serve, drain the poached salmon, place each piece onto a dinner plate and put the hollandaise sauce on the salmon.

Unit 5
The Larder Chef

A. Fill in the blanks and make changes where necessary.

| incorporate into | taste wonderful | egg yolk | lemon juice |

1. The turkey _____.
2. The _____ tastes too sour.
3. The egg consists of _____, egg white and egg crust.
4. The small company is _____ a big company.

B. Choose the ingredients for the poached salmon with hollandaise sauce according to the dialogue.

Ingredients: _____

a. lemon juice   b. shrimp   c. olive oil   d. pepper   e. butter

f. preserved egg   g. salmon fillets   h. crab   i. egg yolk   j. salt

Role-play

Make a simple dialogue between Aida and Leo about how to make the poached salmon with hollandaise sauce.

Reading

Preparing the whole fish **involves scaling**[1], chopping off the **fins**, taking off the **gills**, **gutting** and washing. If the recipe calls for an original shape of fish, you should first gut it by cutting along the **spine**. Then cut parallel to[2] the spine almost up to the top and

85

separate the flesh from the top and bottom of the center bone. Lift out the center bone and small side bones and cut the spine away at the head and tail. Finally, wash the **cavity** and the outside and **arrange** the fish as closely as possible[3] in its original shape.

A. Translate the following sentences into Chinese.

1. I asked the butcher to chop off a piece of meat for me.
_____

2. Raw meat must be kept separate from cooked meat.
_____

3. Different positions call for different skills.
_____

4. Why do you take off all the pictures?
_____

B. Learn the important phrases in the text.

1. involve doing 包含
- Preparing the whole fish involves scaling, chopping off the fins, taking off the gills, gutting and washing. 切鱼包含去鳞、切下鱼翅、取下鱼鳃、取出鱼内脏并清洗干净。

练习：家务事包含做饭、洗盘子、打扫卫生。
英文：_____

2. parallel to 平行
- Then cut parallel to the spine almost up to the top and separate the flesh from the top and bottom of the center bone. 然后与脊椎平行切下直至头部，将鱼肉自头尾从中间向两侧分开。

练习：该公路与铁路平行。
英文：_____

3. as...as possible 尽可能……地
- Finally, wash the cavity and the outside and arrange the fish as closely as possible in its original shape. 最后，将腹腔和外部洗净，尽可能将鱼整理成和原先一样的形状。

练习：应该尽快实施这项法规。
提示：implement
英文：_____

Unit 5

The Larder Chef

Task

Group discussion: the following pictures show the major forms of preprocessed fish. Translate the processing instructions into Chinese.

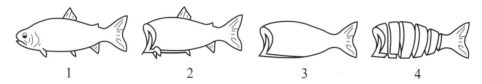

1   2   3   4

1. The whole fish. _____
2. Cut off the head. _____
3. Cut off the fins. _____
4. Gut and slice the whole fish. _____

Further Study

Learn about additional expressions.

English	Chinese	English	Chinese
fin fish	有鳍鱼类	shell fish	贝壳类
fresh water fish	淡水鱼类	conch	海螺
saltwater fish	咸水鱼类	abalone	鲍鱼
catfish	鲇鱼	clam	蛤蜊
trout	鳟鱼	scallop	扇贝
perch	鲈鱼	squid	鱿鱼
carp	鲤鱼	sea slug	海参
tuna	金枪鱼	crayfish	淡水鳌虾
flatfish	比目鱼	crawfish	小龙虾
shark	鲨鱼	cuttlefish	墨鱼

Unit 6

The Executive Level

行政组厨师

Lesson 1
The Line Cook
三厨

Goal

Learn the terms of food ingredients.

Warm-up

Write the correct term below each picture.

| mince | dice | crush | slice | roll cut | julienne |

1. _____
2. _____
3. _____
4. _____
5. _____
6. _____

Unit 6
The Executive Level

Dialogue

Jackie（J）是餐厅的三厨，他在教 Leo（L）切各类蔬菜。

L: Jackie, can you teach me how to cut vegetables?

J: There are different ways to cut various kinds of vegetables.

L: What about the potato?

J: The potato is a kind of **oval** vegetable. So first we have to cut it into half that you will have a flat surface. And cut half-moon panels from the potato halves. We can use the slices as they are at this point.

L: If I want to **julienne** the potato, what can we do next?

J: Stack up the potatoes and move the knife down the row, keeping the tip in contact with the cutting surface to produce juliennes.

L: What about the Chinese cabbage?

J: That's the vegetable with leaves. We begin by **stacking** the leaves with the largest at the bottom, and roll the stack of leaves tightly and slice the Chinese cabbage.

L: Great. And the carrot?

J: That's a long vegetable. Since it is too long to manage, we can cut it into 2 or 3 sections. Cut a thin slice off each carrot section so that it will lie flat. And then cut panels off each carrot section, slicing as thinly as you can.

L: OK. Thanks. I've learned a lot from you.

A. Translate the following sentences into Chinese.

1. He stacks up all his books on the desk.

2. Are you in contact with him?

3. The fish is at the bottom of the lake.

___

B. Describe each step under the corresponding picture according to the dialogue.

★ How to julienne oval vegetables.

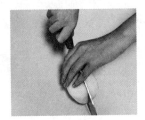  

Step 1: (cut; flat surface)　Step 2: (half-moon panel; slice)　Step 3: (stack; julienne)

★ How to cut vegetables with leaves.

Step 1: (stack; at the bottom)　Step 2: (roll)　Step 3: (slice)

★ How to cut long vegetables.

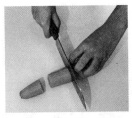 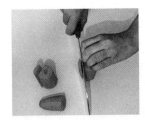 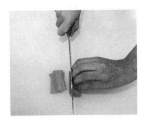

Step 1: (cut; section)　Step 2: (cut; panel)　Step 3: (slice)

Role-play

Make a simple dialogue between Jackie and Leo about how to cut vegetables.

Unit 6

The Executive Level

Reading

Meat is prepared in many ways, as **steaks** or **dried** meat like beef **jerky**. It may be **ground** then **formed** into **patties** (as hamburgers). Some meat is **cured** by smoking, preserving in salt or brine. Other kinds of meat are marinated and barbecued, or simply boiled, roasted or fried. Meat is generally eaten cooked, but there are many traditional recipes that call for raw meat.

Meat is a typical base for[1] making sandwiches. Popular **varieties** of sandwich meat include ham, pork, and other sausages, as well as[2] beef, such as steak and roast beef. Meat can also be canned.

A. Match the English with the Chinese.

1. Would you like some beef jerky?　　　　　a. 女孩子们排成3排。
2. The girls are formed into 3 lines.　　　　　b. 他有各种兴趣爱好。
3. He has varieties of interests.　　　　　　　c. 你要来点牛肉干吗？

B. Learn the important phrases in the text.

1. a base for... ……的基础
- Meat is a typical base for making sandwiches. 肉类是做三明治的一种典型底料。

练习：她把自己一家人的经历作为小说的素材。
提示：experience, materials
英文：_____

2. as well as 和
- Popular varieties of sandwich meat include ham, pork, and other sausages, as well as beef, such as steak and roast beef. 三明治中受欢迎的肉类包含火腿、猪肉、其他香肠和牛肉，如牛排、烤牛肉。

练习：他种菜和花。
英文：_____

Task

Group discussion: classify the following food into categories according to their sources.

chicken wings   pork   sausage   bacon

steak   fresh grade leg   ox tail   pork burger

Categories:
1. From chicken: _____
2. From pig: _____
3. From cattle: _____

Further Study

Learn to be a qualified line cook.

Because communication is important during food service, a line cook must be able to work well with other line cooks while preparing quality food and handling complaints from customers or staff（员工）. The ability to coordinate（协调）several different orders at the same time is also a good skill for a line cook to possess（具备）. If you want to be a good line cook, paying attention to details and working quickly under pressure is essential.

Unit 6
The Executive Level

Lesson 2
The Sous Chef
副总厨

Goal

Learn the ways of storing and preserving foods.

Warm-up

Write the correct term below each picture.

| sun drying | canning | smoking |
| freezing | vacuum packing | salting |

1. _____   2. _____   3. _____

4. _____ 5. _____ 6. _____

Dialogue

Sabina（S）是餐厅的副总厨，她在教 Leo（L）一种保存食物的好方法——熏烤法。

L: Sabina, can you teach me how to **preserve** food?

S: Well, look, I am smoking fish. Smoking is one of ways to preserve food.

L: You are really skillful. How can you smoke fish?

S: I've already bought some fish that was quickly frozen. Now, **defrost** it, clean it, remove the head, tail and fin and wash it in clean water.

L: OK. So what are you doing now?

S: I'm preparing salt-water **brine** and I'll place the fish in the brine for 15 minutes.

(*15 minutes later.*)

S: It's time for us to remove the fish from the brine and rinse it with cold water.

L: Let me help you place the fish on oiled rack. I always watch my mum doing this.

S: Thank you. So we just need to keep the temperature low to around 65 ℃ and increase heat after the first 2 hours to around 95 ℃.

L: It's easy and we continue smoking until the fish is flaky and cooked through, right?

S: Smart!

Unit 6
The Executive Level

A. Reorder the letters.

Letters	Words
a r k c	r
o r e n z f	f
e n i r b	b
l a f y k	f

B. Fill in the blanks and make changes where necessary.

> smoked fish   be skillful   remove...from...   continue reading

1. He _____ in making some crafts.
2. Do you like _____?
3. I want to stay at home and _____ my cooking book.
4. She _____ mud _____ her shoes.

Role-play

Make a simple dialogue between Sabina and Leo about how to preserve food.

Reading

Food **preservation** is the **process** of **treating** and handling food to stop or greatly slow down spoilage. The methods of preserving include drying, freeze drying, **vacuum**-packing[1], and **canning**.

The traditional way to preserve garden gains[2] is canning. This involves heating up both **jars** and produce to kill the **germs** and then sealing the produce within the jars.

The most highly **reliable** ancient method for preserving food is drying. Since before recorded history people have dried herbs, meats, fruits and vegetables to store them for use at a later date.

There is a new and popular way to preserve food that has just caught on in the last decade or so. Vacuum sealing is a method whereby food is put inside a heavy plastic wrap and all the airs removed; then it is sealed.

A. Match the English with the Chinese.

(　　) The machine slowed down and stopped.
(　　) The iron heats up in the sun.
(　　) When did the recorded history of Britain begin?
(　　) Price has risen steadily in the past decade.

1. 机器减慢速度，最后停了下来。
2. 英国有记载的历史是从什么时候开始的？
3. 过去10年间物价稳步上涨。
4. 铁在太阳下变热。

B. Learn the important word and phrase in the text.

1. vacuum-packing 真空包装
- The methods of preserving include drying, freeze drying, vacuum-packing, and canning. 保存的方法包括烘干、冷冻干燥、真空包装和装罐保存。

练习：目前咸鸭蛋的包装规模化生产一般采用真空包装、高温杀菌工艺。
提示：salted, sterilization
英文：_____

2. the way to do sth. 做某事的方法
- The traditional way to preserve garden gains is canning. 保存果蔬食品的传统做法是装罐保存。

练习：用什么方法能把这个弄干净？
英文：_____

Task

Read the text again and circle Yes or No.

1. Canning is the most highly reliable ancient method for preserving food.　　Yes　　No
2. One of the traditional ways to preserve garden gains is drying.　　Yes　　No
3. The methods of preserving include vacuum-packing, freeze drying, drying and canning.　　Yes　　No

Unit 6
The Executive Level

Further Study

Learn to be a qualified sous chef.

A sous chef is the second in command in a kitchen. In French, the term literally means "under chef". The job is challenging, but extremely rewarding（值得做的）when a kitchen runs smoothly and pleasantly. The sous chef is on the ground in the kitchen every day, keeping track of a wide range of issues（处理各项事务）and working to ensure the food in the restaurant is of the highest quality. A sous chef certainly does some cooking, but the job is about much more than just preparing food. Sous chefs supervise food preparation and safety all over the kitchen, keeping an eye on the staff and ensuring that the dishes stay true to the vision of the executive chef.

Lesson 3
The Executive Chef
行政总厨

Goal

Learn to succeed in a job interview.

Warm-up

Write the correct term below each picture.

| interview | apply for a job | catering industry |

1. _____

2. _____

3. _____

Unit 6
The Executive Level

Dialogue

Howard（H）即将毕业，他在接受餐厅行政总厨Smith（S）的面试。

H: Good morning Mr. Smith. My name is Howard, Leo's friend.

S: Good morning, Howard. Please take a seat. Leo has recommended you to us and he speaks highly of you. You are  applying for the cook in our restaurant. Can you introduce yourself first?

H: Yes. Thank you for giving me this opportunity to have this interview. I graduated from ABC School, majoring in culinary art. I was awarded the first prize of Cooking Contest for East Chinese Vocational Students.

S: Very **impressive resume**. Have you taken any course on professional English?

H: Yes. I had the course of English for culinary art in my second year in the school and I took a part-time job in Apple Bees, an American restaurant during summer vacation.

S: Can you describe your **personality**?

H: OK. I am an easy-going person and I like team work. I am patient and **energetic**. I can work very hard and I am a quick learner.

S: Sounds great. Then, what are your disadvantages?

H: Well, I haven't had any full-time job, so I need to gain more experience.

S: OK. That's all. Thank you for your time, Howard. You'll take a basic cooking test, if you pass the interview. And we will inform you as soon as possible.

H: Thank you very much Mr. Smith. Good-bye.

A. Match the English with the Chinese.

(   ) A CV is needed when you apply for a job.
(   ) He majors in culinary art.
(   ) The executive chef is good at culinary art.
(   ) Thank you for giving me this opportunity to make a speech here.

1. 这位行政总厨擅长烹饪艺术。
2. 谢谢你给我这次演讲的机会。
3. 求职的时候需要有份简历。
4. 他主修烹饪艺术。

B. Tips for a job interview.

Take a copy of your CV with you.

Wear suitable interview clothes.

Arrive on time for your job interview.

Do not smoke, chew gum, or eat garlic beforehand.

Have a good firm handshake.

Wait until you are offered a chair before you sit down.

Unit 6
The Executive Level

Role-play

Make a simple dialogue between Smith and Howard about the interview.

Reading

When planning a **menu**, you should **design courses** that interact and blend harmoniously with the next course. Remember to imagine the **impact** and final **impression**.

For instance, if you plan a **luncheon** with Italian sausages and mustard as the main course, the courses before and following should be light such as simple salads and juicy **flavorful** fruits. This will cool the stomach and prevent appetite overload.

Appetizers: If you are serving appetizers only, figure on 6 to 8 pieces per adult. If the appetizers are before a dinner or buffet service, 4 to 6 pieces should suffice. It is preferable to offer a smaller variety[1] but each in greater quantities[2]. This saves time and costs. Also the amount of appetizers may depend on what it is.

A. Write the correct term below each picture according to the reading material.

1. 开胃菜_____   2. 自助餐_____   3. 香肠_____

4. 午餐会_____   5. 菜单_____   6. 芥末_____

103

B. Learn the important phrases in the text.

1. preferable to do sth. 更可取的是
- It is preferable to offer a smaller variety... 更可取的是品种少……

练习：有健康而无财富比有财富而无健康更可取。

提示：wealth, health

英文：_____

2. in greater quantities 大量，数量多
- ...but each in greater quantities. ……但数量多。

练习：他们送来大量豆腐、白糖和其他易于吸收的食物。

英文：_____

Task

Translate the following sentences into Chinese.

1. Water never blends with oil.
_____

2. The main course comes after the appetizer.
_____

3. The buffet service is very convenient.
_____

4. It depends on the weather.
_____

Further Study

Learn to be a qualified executive chef.

The executive chef, who oversees（监督）all kitchen staff, food preparation and cooking activities, is the key to a restaurant's success. Most of his time is spent outside the kitchen researching on trends（趋势）in the food industry, planning and writing menus, budgeting and financial planning, and participating in business development. The executive chef also recruits and hires staff, supervises their activities and instructs（指导）cooks in preparation, cooking and presentation of food.

Appendix 1

Grammar in Use

简单句

简单句是英语句子结构的基础。在英语中，有以下几种简单句。

主谓。结构：主语 + 谓语。表示主语的动作或状态。例如：

> **We work** together in the kitchen.
> 我们一起在厨房工作。

主谓宾。结构：主语 + 谓语 + 宾语。表示主语对宾语的动作或状态。例如：

> **She oversees** the daily **operations** in the kitchen.
> 她负责监督厨房的日常运营。

主系表。结构：主语 + 系动词 + 表语。描述主语的状态或特征。例如：

> **I'm** really **excited** to be here.
> 能来到这里我真的很激动。

主谓双宾。结构：主语 + 谓语 + 间接宾语（人）+ 直接宾语（物）。例如：

> Can **you give me** some **tips**?
> 你能指点指点我吗？

主谓宾补。结构：主语＋谓语＋宾语＋宾语补足语。宾语补足语补充说明宾语的状态或特征。例如：

> **You'll find it organized and efficient.**
> 你会发现它井井有条，效率很高。

形容词：作定语

英语形容词作定语，主要用于修饰名词，说明其性质、特征或状态。

通常，形容词位于被修饰的名词之前，如 **clean** water、**kitchen** assistant。

当形容词短语、表语形容词或某些特殊结构的形容词修饰名词时，它们可能后置，如 instructions **for use**。

此外，多个形容词作定语时，应遵循以下顺序：

1. 限定词。如冠词 the、a，物主代词 my、his、her 和指示代词 this、that 等。
2. 数词。如序数词 first、second 和基数词 one、two 等。
3. 描绘性形容词。如 delicious、heavy 等，用于描述名词的抽象性质或特征。
4. 大小、形状、年龄、新旧。如 large、round、new 等，用于描述名词的具体形态或状态。
5. 颜色。如 red、blue、green 等，用于描述名词的颜色。
6. 国籍。如 Chinese、French、Pakistan 等，用于表示名词的来源或国籍。
7. 材料。如 wooden、plastic 等，用于描述名词的构成材料。
8. 用途或类别。如 **pastry** tool，用于说明名词的用途或所属类别。

遵循以上顺序有助于保持语句的逻辑性。例如：

> **I usually use a clean heavy flat stone.**
> 我一般用一块干净、平整且比较沉的石块。

副词：作状语

副词在句子中主要用作状语，用来修饰动词、形容词、其他副词或全句，表达时间、地点、程度等概念。副词作为状语，可以使句子的表达更加完整和准确。

时间副词，如 now、soon、ago 等，表达动作发生的时间。它们通常在句尾，但在强调时也可放在句首。例如：

> **Now add all of the salt and sugar.**
> 现在加入所有的盐和糖。

地点副词，如 here、there、abroad 等，表达动作发生的地点。它们通常在句尾，紧跟在动词或宾语之后。例如：

> Thank you for giving me this opportunity to make a speech **here**.
> 谢谢你给我这次演讲的机会。

程度副词，如 very、much、almost 等，用来修饰形容词或副词，表达动作或状态的程度。它们通常在被修饰词之前。例如：

> Among many types of foods, beans are perhaps **almost** the perfect type of food.
> 在众多食物中，豆类几乎是完美的食物。

介词

介词是一种虚词，通常位于名词或代词之前，表示这些词与句中其他成分的关系。

介词之后一般跟有名词、代词或相当于名词的其他词类、短语或从句，在句中可作状语、表语、补语、定语等，起到连接和说明的作用。

介词有多种分类，如地点介词 in、on、under 等，时间介词 in、on、at 等，方式介词 by、with、through 等，原因介词 for、(because) of 等。介词的具体用法取决于它表达的关系。

in 可以表示在某地点范围内，也可以表示在某时间范围内。例如：

> Steamed buns, or *baozi*, are a kind of staple food **in** China for breakfast or any other time **in** a day.
> 在中国，包子是一种主食，可在早餐或一天中的任何时间食用。

with 则表示使用某种工具或手段。例如：

> Outline fireplace/chimney area **with** a round decorating tip.
> 用圆形挤花嘴画出壁炉或烟囱的外形。

并列句

并列句是由两个或两个以上单词、短语或简单句，通过并列连词连接而成的句子结构。

常见的并列连词有 and、but、or、both...and...、not only...but also...、as well as、neither...nor...、either...or... 等，它们不仅连接并列句子成分，还表明其间的逻辑关系。

and 可以连接两个简单句，表明两者之间的并列关系。例如：

Leo has recommended you to us **and** he speaks highly of you.
利奥向我们推荐你,他对你评价很高。

but 表示转折关系。例如:

A sous chef certainly does some cooking, **but** the job is about much more than just preparing food.
副主厨确实会参与烹饪工作,但他们的职责远不止准备食物那么简单。

or 表示选择关系。例如:

What kind of pans do you want, saute pans **or** frying pans?
您需要什么样的锅,煎炒锅还是炸锅?

not only...but also... 表示递进关系。例如:

They **not only** prevent certain diseases, **but also** reverse them.
它们不仅可以预防某些疾病,还可以祛除疾病。

复合句

复合句是由一个主句和一个或多个从句,通过关联词连接而成的句子结构。

主句是复合句的核心部分,它本身可以独立成句,表达完整的意思。从句则依附于主句,作为主句的一个成分,如主语、宾语、表语、定语、状语等,用以补充额外的信息或细节。

从句根据其功能可分为名词性从句(如主语从句、宾语从句、表语从句、同位语从句)、形容词性从句(定语从句)和副词性从句(状语从句)。每种从句都有其特定的引导词和语法规则。

that 可以引导定语从句,补充主句中额外的信息或细节。例如:

People usually come with savory filling **that** is added to the dough before steaming.
通常在蒸包子前,人们会在面团里加些美味的馅料。

地点状语从句

地点状语从句是英语语法中一种常见的从句,主要用来修饰主句中的动作或状态发生的地点。

地点状语从句通常由 where、wherever、anywhere、everywhere 等引导词引导。例如：

> Pinch the 2 places **where the wrapper is folded over**.
> 将皮重叠的两边捏住。

原因状语从句

原因状语从句是指在复合句中用于说明主句动作发生原因的从句。

常见的原因状语从句引导词有 because、since、as、for 等，以及复合引导词 now that、seeing that、for the reason that 等。

其中，because 语气最强。例如：

> **Because communication is important during food service**,
> a line cook must be able to work well with other line cooks while preparing quality food
> and handling complaints from customers or staff.
> 因为餐饮服务中的沟通至关重要，
> 一名三厨在准备高质量食物的同时，必须能够与其他三厨很好地配合，
> 并妥善处理来自顾客或员工的投诉。

since 和 as 语气较弱，表示已知或明显的理由；for 语气更弱，常用于引出补充性原因。例如：

> **Since it is too long to manage**, we can cut it into 2 or 3 sections.
> 因为它太长了，我们可以把它切成 2~3 段。

now that 通常用于引导已知的事实原因；seeing that 是较为正式的用法，用于引导显而易见或已被观察到的原因；for the reason that 明确指出具体原因，常见于正式写作或演讲中。

结果状语从句

结果状语从句是用于补充说明主句中谓语动词发生结果的从句，通常位于主句之后。其常用引导词或短语包括 so that、so...that...、such...that... 等。

so that 表示"结果"或"所以"。例如：

> Cut a thin slice off each carrot section **so that it will lie flat**.
> 每段胡萝卜切下一片薄片，这样就可以放平了。

so...that... 表示"如此……以至于……",其中 so 后接形容词或副词。例如:

> Beans, in fact, are **so** filled with nutrition and **so** reduced in calories **that, if you happen to be dieting, beans are the right food for you**.
> 事实上,豆类不但营养丰富,而且低卡路里。若你正巧在节食,豆类正是适合你的食品。

such...that... 也表示"如此……以至于……",但 such 后接名词或名词词组。

方式状语从句

方式状语从句说明主句中动作发生的方式或手段,其常用的引导词有 as、as if、as though、like 等。as 意思是"如同……""按照……的方式"。as if 和 as though 都表示"好像""仿佛"。like 表示"像……一样"。例如:

> Meat is prepared in many ways, as steaks or dried **meat like beef jerky**.
> 肉类有多种烹调法,如牛排或干肉类像牛肉干。

条件状语从句

条件状语从句是用来表示主句动作发生所需条件的从句。

条件状语从句常由 if、unless、as/so long as、only if、in case、provided/providing (that)、on condition that 等引导。例如:

> **If you want to be a good line cook,**
> paying attention to details and working quickly under pressure is essential.
> 如果你想成为一名优秀的三厨,注重细节并在压力下快速工作是非常重要的。

让步状语从句

让步状语从句主要用于表达在某个条件或情况下,主句所描述的动作或状态仍然会发生或存在。

常见的引导词有 although、though、even if、even though、whether...or...、no matter+ 疑问词以及疑问词 -ever 等。需要注意的是,although 和 though 不能与 but 连用,但可以与 still、yet 等词连用。

例如：

> **Although it looks like a dish simply made of shrimps,**
> the material's processing and cooking can be quite complicated.
> 虽然这道菜看起来是用虾仁简易制成的，但食材处理和烹制的过程相当复杂。

定语从句：which v.s. that

定语从句是英语中一种重要的语法结构，用于修饰主句中的名词或代词。

定语从句由关系代词（如 which、that、who、whom、whose）或关系副词（如 when、where、why）引导。关系代词用于指代人或物，关系副词则用于表示时间、地点或原因，在从句中作状语。其中，which 在非正式场合和口语中更为常见，而 that 则更多地用于正式场合或书面语中。which 通常用于引导非限定性定语从句，即从句对先行词进行额外说明，但去掉从句后，句子的主要意思不受影响。which 在从句中可以作主语、宾语或表语，它与主句往往用逗号隔开。例如：

> It has a more delicate flavor and texture than mutton, **which** takes on a rich gamy flavor.
> 这种肉味道、肉质比普通羊肉鲜嫩。普通羊肉通常有腥味。

that 则用于引导限定性定语从句，即从句对先行词进行必要的限定，去掉后句子意思不完整或发生改变。that 在从句中同样可以作主语、宾语或表语，但无逗号与主句隔开。例如：

> When planning a menu, you should design courses
> **that** interact and blend harmoniously with the next course.
> 在设计菜单时要注意前后菜肴的和谐统一。

定语从句：who v.s. whom

who 和 whom 都能用于引导定语从句，指代前面提到的"人"。

两者的主要区别在于，who 在定语从句中作主语，而 whom 在定语从句中作宾语。例如：

> The executive chef, **who oversees all kitchen staff, food preparation and cooking activities**, is the key to a restaurant's success.
> 行政主厨是餐厅成功的关键，他们负责监督所有厨房员工、食品准备及烹饪活动。

主语从句

主语从句是英语中一种重要的从句类型，它在句子中担任主语的角色。主语从句可由多种引导词引导，如连词 that（无实际意义，但不能省略）、whether（表示选择或疑问）、连接代词（如 who、what、which 等）和连接副词（如 when、where、how 等）。例如：

> **What makes the dish good** is that each bite first gives crisp feeling at the skin of the shrimps, but when you bite further you will soon feel the tender and juicy from the shrimps' body.
> 这道菜的精华在于一入口尝到的是虾外皮的酥脆感，
> 再进一步品尝会发现虾仁柔嫩多汁。

如果主语从句较长，为了避免头重脚轻，通常会在句首使用形式主语 it 来代替主语，而真正的主语从句则置于句末。

宾语从句

在复合句中，充当宾语的从句称为宾语从句。它通常位于及物动词、介词或复合谓语之后，由从属连词（如 that、if、whether）或连接代词和连接副词（如 who、what、when、why、how）等引导。

that 引导陈述句宾语从句时可省略。例如：

> **Some people say (that) this dish is the best dish of Shanghai**.
> 有人说这是上海最好的一道菜。

连接代词和连接副词则根据从句的具体需要选择，作不同的句子成分。例如：

> Think about how many times you cut onions each week and you will understand **why it is so important you learn how to do it safely and properly**.
> 只要想想你每周要切多少次洋葱，你就明白学会安全正确地切洋葱为何如此重要了。

表语从句

表语从句是在复合句中用作表语的从句，它位于主句中的 be、remain、become 等之后。

Appendix 1
Grammar in Use

引导词包括连词（如 that、whether）、连接代词（如 who、what、which）和连接副词（如 when、where、why、how）等。that 在引导表语从句时通常不翻译且一般不省略，但在口语或非正式文体中有时可省略。

例如：

> After I took my first baking class, I knew it was **what I really wanted to do**.
> 第一次上烘焙课后，我就意识到这是我真正想做的事。

Appendix 2

Words and Expressions

Pre-Unit

announcement [əˈnaunsmənt] *n.* 宣布，宣告
graduate [ˈgrædʒu,eɪt] *v.* (从……) 毕业
oversee [,əuvəˈsi:] *v.* 监督，管理
operation [,ɒpəˈreɪʃən] *n.* 运营
universal [ju:nɪˈvɜ:s(ə)l] *adj.* 通用的
flavor [ˈfleɪvə(r)] *n.* 味道
culture [ˈkʌltʃə(r)] *n.* 文化
tradition [trəˈdɪʃən] *n.* 传统
unique [juˈni:k] *adj.* 独特的
innovate [ˈɪnəveɪt] *v.* 创新
experiment [ɪkˈsperɪmənt] *v.* 尝试

not only...but also...   不仅……而且……
all over the world   全世界
enjoy doing sth.   喜欢做某事

Unit 1

Lesson 1

crockery [ˈkrɒkəri] *n.* 陶器
liquid [ˈlɪkwɪd] *n.* 液体
recipe [ˈresəpi] *n.* 食谱
dice [daɪs] *v.* 切丁
slice [slaɪs] *v.* 切片
chop [tʃɒp] *v.* 剁碎
onion [ˈʌnjən] *n.* 洋葱
stem [stem] *n.* 茎
grab [græb] *v.* 抓取
root [ru:t] *n.* 根
attach [əˈtætʃ] *v.* 连接
multiple [ˈmʌltɪpl] *adj.* 多个的
degree [dɪˈgri:] *n.* 度
finger [ˈfɪŋgə] *n.* 手指

pickled meat   腌肉
stew pot   炖锅
at a rolling boil   在沸腾
pickling liquid   腌汁
stem end   根茎端
face up   朝上
cut across   横切

Lesson 2

mustard [ˈmʌstəd] *n.* 黄芥末
clove [kləuv] *n.* 蒜瓣
bake [beɪk] *v.* 烘烤
thermometer [θəˈmɒmɪtə] *n.* 温度计
yolk [jəuk] *n.* 蛋黄

115

vinegar [ˈvinigə] n. 醋
vitamin [ˈvitəmin] n. 维生素
mineral [ˈminərəl] n. 矿物质
appetite [ˈæpitait] n. 胃口
physician [fiˈziʃən] n. 内科医生
disease [diˈzi:z] n. 疾病
reverse [riˈvə:s] v. 祛除
calcium [ˈkælsiəm] n. 钙
iron [ˈaiən] n. 铁
spinach [ˈspinidʒ] n. 菠菜

bread crumb 面包屑
combine...with... 把……与……混合
brown sugar 红糖
suggest (that) sb. (should) do... 建议某人做……
be vital to... 对……至关重要

Lesson 3

unload [ʌnˈləud] v. 从……卸下货物
unpack [ʌnˈpæk] v. 从……取出
cooker [ˈkukə] n. 炊具，锅
scissors [ˈsizəz] n. 剪刀
mince [mins] v. 切碎，剁碎，绞碎
toast [təust] n. 烤面包，吐司
challenging [ˈtʃæləndʒiŋ] adj. 挑战性的
mixer [ˈmiksə] n. 搅拌器，混合器
chemical [ˈkemikəl] n. 化学药品
countertop [ˈkauntətɔp] n. 工作台面
handle [ˈhændl] v. 处理
uncooked [ʌnˈkukt] adj. 生的
coating [ˈkəutiŋ] n. 保护层
moisture [ˈmɔistʃə] n. 水分
spoil [spɔil] v. 变质
rub [rʌb] v. 揉擦
detergent [diˈtədʒənt] n. 洗涤剂
scrub [skrʌb] v. 擦洗

land a job 找到工作
operate...machine 操作……机器

dispose of sth. 处理某事
what's more 除此之外
used-by date 保质期
uncooked meat 生肉
along with... 与……一起
running water 自来水

Unit 2
Lesson 1

pastry [ˈpeistri] n. 面点
profession [prəˈfeʃən] n. 职业
culinary [ˈkju:lənəri] adj. 烹饪的
decoration [dekəˈreiʃən] n. 装饰
flour [ˈflauə(r)] n. 面粉
powder [ˈpaudə(r)] n. 粉末
tart [tɑ:t] n. 蛋挞
dough [dəu] n. 生面团
filling [ˈfiliŋ] n. 馅
croissant [krwɔˈsɑ:nt] n. 羊角面包
characteristic [ˌkæriktəˈristik] adj. 特有的

be interested in... 对……感兴趣
have a passion for... 对……有激情
various kinds of... 各种各样的……
refer to 指代
fat content 脂肪含量

Lesson 2

sesame [ˈsesəmi] n. 芝麻
wrapper [ˈræpə] n. 包装材料（饺子皮）
pinch [pintʃ] v. 捏住
marketplace [ˈmɑ:rkitˈpleis] n. 市场
typical [ˈtipikəl] adj. 典型的
staple [ˈsteipl] adj. 主要的
savory [ˈseivəri] adj. 美味的
tendon [ˈtendən] n. 筋
chewy [ˈtʃu:i] adj. 耐嚼的
spicy [ˈspaisi] adj. 辣的

stinky ['stiŋki] adj. 发臭的
horrible ['hɔrəbl] adj. 恐怖的

prevent sth. from... 防止某物……
dim sum 小吃
add to 加入
be known as... 被称为……
be famous for... 以……而著名

Unit 3
Lesson 1

cucumber ['kju:kʌmbə] n. 黄瓜
mint [mint] n. 薄荷
buttermilk ['bʌtəmilk] n. 白脱牛奶
yogurt ['jəugət] n. 酸奶
platter ['plætə] n. 拼盘
similar ['similə] adj. 类似的
shape [ʃeip] n. 形状
melon ['melən] n. 甜瓜
mango ['mæŋgəu] n. 芒果
fan [fæn] v. 摆成扇形
assemble [ə'sembl] v. 组装
flesh [fleʃ] n. 果肉
pit [pit] n. 果核
equal ['i:kwəl] adj. 相同的
proceed [prə'si:d] v. 进行下去

sound+adj. 听起来……
deal with 处理
continue doing sth. 继续做某事
fruit platter 水果拼盘
begin to do 开始做
slice down 切下
in half 成两半
proceed to 接着做

Lesson 2

gingerbread ['dʒindʒəbred] n. 姜饼

nutmeg [nʌtmeg] n. 肉豆蔻
zest [zest] n. 柑橘外皮
strip [strip] n. 条状物
rinse [rins] v. 冲洗
ramekin ['ræmikin] n. 小盘子
whisk [hwisk] n. 搅拌器
creative [kri'eitiv] adj. 创造（性）的
tip [tip] n. 技巧
icicle ['aisikl] n. 冰柱
pressure ['preʃə] n. 压力
pipe [paip] vt. 裱花
bottom ['bɔtəm] n. 底部
pretzel ['pretsəl] n. 椒盐脆饼
wafer ['weifə] n. 威化饼

set...aside 把……放在一边
store...in the fridge 把……放进冰箱
decorating tip 挤花嘴
from top to bottom 从顶部到底部
side by side 并排

Unit 4
Lesson 1

turnip ['tə:nip] n. 萝卜
flour ['flauə] v. 在……上撒面粉
citrus ['sitrəs] n. 柑橘
cinnamon ['sinəmən] n. 肉桂皮
unmold [ʌn'məuld] v. 把……从模子里取出
preference ['prefərəns] n. 爱好
tough [tʌf] adj.（肉）老的
term [tə:m] n. 术语
raw [rɔ:] adj. 生的
rare [rɛə] adj.（牛排）一分熟
medium ['mi:diəm] adj.（牛排）五分熟
throughout [θru:'aut] prep. 各处
char [tʃɑ:] v. 烤焦
overcook [ˌəuvə'kuk] v. 烘焙过度，煎焦

do sb. a favor 帮某人一个忙

far away from... 离……很远
clarified butter 澄清黄油
blue rare （牛排）近生
medium rare （牛排）三分熟
medium well done （牛排）七分熟
well done （牛排）全熟

Lesson 2

saute ['səutei] v. n. 煎，炒
flat [flæt] adj. 平坦的
dredge [dredʒ] v.（用面粉、糖等）撒
dip [dip] v. 浸
parsley ['pɑːsli:] n. 西芹
grated ['greitid] adj. 磨碎的
crystal ['kristəl] adj. 水晶的
shrimp [ʃrimp] n. 虾肉
material [mə'tiəriəl] n. 材料（食材）
process ['prəuses] v. 处理
complicated ['kɔmplikeid] adj. 复杂的
bite [bait] n. 一口
crisp [krisp] adj. 酥脆的
further ['fə:ðə] adv. 进一步
juicy ['dʒu:si] adj. 多汁的

kitchenware store 厨具店
saute pan 煎炒锅
frying pan 炸锅
be popular among... 受……欢迎
crystal shrimp 水晶虾仁

Lesson 3

lace [leis] v. 系牢，系紧
degrease [di:'gri:z] v. 除去……的脂肪
doneness ['dʌnnis] n.（肉）的熟度
roast [rəust] adj. 烤的
famous ['feiməs] adj. 著名的
imperial [im'piəriəl] adj. 帝国的
era ['iərə] n. 时代
prize [praiz] v. 珍视

crispy ['krispi] adj. 脆的
household ['haushəuld] adj. 家喻户晓的

on both sides 两侧都
Peking Roast Duck 北京烤鸭
be considered as sb./sth. 被认为是……
be prized for sth. 因……而受到珍视
household name 家喻户晓的名字
hoisin sauce 海鲜酱
sweet bean sauce 甜面酱

Lesson 4

crush [krʌʃ] v. 压碎
cancer ['kænsə] n. 癌
peel [pi:l] v. 剥皮
evaporate [i'væpəreit] v.（使）蒸发
bean [bi:n] n. 豆
nutrition [nju:'triʃən] n. 营养
calorie ['kæləri] n. 卡路里
diet ['daiət] v. 规定饮食
source [sɔ:s] n. 来源
fiber ['faibə] n. 纤维素
dietary ['daiətəri] adj. 饮食的
effect [i'fekt] n. 效果

season...with... 用……给……调味
in fact 事实上
be filled with... 充满……
happen to 正巧
a source of... ……的一个来源

Unit 5

Lesson 1

bone [bəun] v. 剔骨
abattoir ['æbətwɑ:] n. 屠宰场
saw [sɔ:] n. 锯
apprenticeship [ə'prentiʃip] n. 学徒身份
cut [kʌt] n. 切下的肉块

tender ['tendə] *adj.* 嫩的
muscle ['mʌsl] *n.* 肌肉
texture ['tekstʃə] *n.* 质地
outer ['autə] *adj.* 外表的
pinkish ['piŋkiʃ] *adj.* 浅粉色的
lamb [læm] *n.* 羔羊肉
gamy ['geimi] *adj.* 有腥味的

be responsible for... 对……负责
be self-taught 自学的
bone out 去骨
start doing sth. 开始做某事
part-time job 兼职
sweep the floor 扫地
cherry red 樱桃红
take on 显现

Lesson 2

poach [pəutʃ] *v.* 水煮
salmon ['sæmən] *n.* 三文鱼
fillet ['filit] *n.* 去骨鱼片
thicken ['θikən] *v.* （使）变稠
incorporate [in'kɔ:pəreit] *v.* 并入
involve [in'vɔlv] *v.* 包含
scale [skeil] *v.* 去鳞
fin [fin] *n.* 鱼翅
gill [dʒil] *n.* 鱼鳃
gut [gʌt] *v.* 取出……的内脏
spine [spain] *n.* 脊椎
separate ['sepəreit] *v.* 分开
cavity ['kæviti] *n.* 腔
arrange [ə'reindʒ] *v.* 排列

taste wonderful 味道非常好
lemon juice 柠檬汁
egg yolk 蛋黄
incorporate into 并入
chop off 切下
take off 取下

call for 要求

Unit 6
Lesson 1

oval ['əuvəl] *adj.* 椭圆形的
julienne [dʒu:li'en] *v.* 把（食物）切成细丝
stack [stæk] *v.* 使成叠地放在……
steak [steik] *n.* 牛排
dried [draid] *adj.* 干的
jerky ['dʒə:ki] *n.* 肉干
grind [graind] *v.* 碾磨
form [fɔ:m] *v.* 成形
patty ['pæti] *n.* 肉饼
cure [kjuə] *v.* 加工处理
variety [və'raiəti] *n.* 种类，各种

cut...from... 从……切……
stack up 叠放
in contact with 接触
at the bottom 在底部
beef jerky 牛肉干
be formed into... 成为……状
varieties of 各种

Lesson 2

preserve [pri'zə:v] *v.* 保存
defrost [di:'frɔst] *v.* 解冻
brine [brain] *n.* 盐水
preservation [,prezə'veiʃən] *n.* 保存
process ['prəuses] *n.* 过程
treat [tri:t] *v.* 处理
vacuum ['vækjuəm] *n.* 真空
can [kæn] *v.* 把（食品）装罐保存
jar [dʒɑ:] *n.* 罐
germ [dʒə:m] *n.* 细菌
reliable [ri'laiəbl] *adj.* 可靠的

smoke fish 做熏鱼

be skillful　熟练
remove...from...　从……中移出……
slow down　减慢速度
heat up　加热
recorded history　有记载的历史
in the...decade　在……10年间

Lesson 3

impressive [im'presiv] *adj.* 令人赞叹的
resume ['rezjumei] *n.* 简历
personality [pɔːsə'næliti] *n.* 性格
energetic [enə'dʒetik] *adj.* 精力充沛的
menu ['menjuː] *n.* 菜单
design [di'zain] *v.* 设计
course [kɔːs] *n.* 一道菜

impact ['impækt] *n.* 影响
impression [im'preʃən] *n.* 印象
luncheon ['lʌntʃən] *n.* 午餐会
flavorful ['fleivəful] *adj.* 可口的
appetizer ['æpitaizə] *n.* 开胃品

apply for　申请
give sb. the opportunity to do...　给某人做……的机会
major in　主修
culinary art　烹饪艺术
blend with...　与……融合
main course　主菜
buffet service　自助餐服务
depend on　取决于，依靠

Appendix 3

Vocabulary

A

a source of... ……的一个来源	4-4
abattoir ['æbətwɑ:] n. 屠宰场	5-1
add to 加入	2-2
all over the world 全世界	P-U
along with... 与……一起	1-3
announcement [ə'naunsmənt] n. 宣布, 宣告	P-U
appetite ['æpitait] n. 胃口	1-2
appetizer ['æpitaizə] n. 开胃品	6-3
apply for 申请	6-3
apprenticeship [ə'prentiʃip] n. 学徒身份	5-1
arrange [ə'reindʒ] v. 排列	5-2
assemble [ə'sembl] v. 组装	3-1
at a rolling boil 在沸腾	1-1
at the bottom 在底部	6-1
attach [ə'tætʃ] v. 连接	1-1

B

bake [beik] v. 烘烤	1-2
be considered as sb./sth. 被认为是……	4-3
be famous for... 以……而著名	2-2
be filled with... 充满……	4-4
be formed into... 成为……状	6-1

be interested in... 对……感兴趣	2-1
be known as... 被称为……	2-2
be popular among... 受……欢迎	4-2
be prized for sth. 因……而受到珍视	4-3
be responsible for... 对……负责	5-1
be self-taught 自学的	5-1
be skillful 熟练	6-2
be vital to... 对……至关重要	1-2
bean [bi:n] n. 豆	4-4
beef jerky 牛肉干	6-1
begin to do 开始做	3-1
bite [bait] n. 一口	4-2
blend with... 与……融合	6-3
blue rare （牛排）近生	4-1
bone [bəun] v. 剔骨	5-1
bone out 去骨	5-1
bottom ['bɔtəm] n. 底部	3-2
bread crumb 面包屑	1-2
brine [brain] n. 盐水	6-2
brown sugar 红糖	1-2
buffet service 自助餐服务	6-3
buttermilk ['bʌtəmilk] n. 白脱牛奶	3-1

C

| calcium ['kælsiəm] n. 钙 | 1-2 |

121

call for 要求	5-2
calorie ['kæləri] n. 卡路里	4-4
can [kæn] v. 把（食品）装罐保存	6-2
cancer ['kænsə] n. 癌	4-4
cavity ['kæviti] n. 腔	5-2
challenging ['tʃæləndʒɪŋ] adj. 挑战性的	1-3
char [tʃɑː] v. 烤焦	4-1
characteristic [ˌkærɪktə'rɪstɪk] adj. 特有的	2-1
chemical ['kemɪkəl] n. 化学药品	1-3
cherry red 樱桃红	5-1
chewy ['tʃuːi] adj. 耐嚼的	2-2
chop [tʃɔp] v. 剁碎	1-1
chop off 切下	5-2
cinnamon ['sɪnəmən] n. 肉桂皮	4-1
citrus ['sɪtrəs] n. 柑橘	4-1
clarified butter 澄清黄油	4-1
clove [kləʊv] n. 蒜瓣	1-2
coating ['kəʊtɪŋ] n. 保护层	1-3
combine...with... 把……与……混合	1-2
complicated ['kɔmplɪkeɪd] adj. 复杂的	4-2
continue doing sth. 继续做某事	3-1
cooker ['kʊkə] n. 炊具，锅	1-3
countertop ['kaʊntətɔp] n. 工作台面	1-3
course [kɔːs] n. 一道菜	6-3
creative [krɪ'eɪtɪv] adj. 创造（性）的	3-2
crisp [krɪsp] adj. 酥脆的	4-2
crispy ['krɪspi] adj. 脆的	4-3
crockery ['krɔkəri] n. 陶器	1-1
croissant [krwɑː'sɑːnt] n. 羊角面包	2-1
crush [krʌʃ] v. 压碎	4-4
crystal ['krɪstəl] adj. 水晶的	4-2
crystal shrimp 水晶虾仁	4-2
cucumber ['kjuːkʌmbə] n. 黄瓜	3-1
culinary ['kjuːlɪnəri] adj. 烹饪的	2-1
culinary art 烹饪艺术	6-3
culture ['kʌltʃə(r)] n. 文化	P-U
cure [kjʊə] v. 加工处理	6-1
cut [kʌt] n. 切下的肉块	5-1
cut across 横切	1-1
cut...from... 从……切成……	6-1

D

deal with 处理	3-1
decorating tip 挤花嘴	3-2
decoration [dekə'reɪʃən] n. 装饰	2-1
defrost [diː'frɔst] v. 解冻	6-2
degrease [diː'griːz] v. 除去……的脂肪	4-3
degree [dɪ'griː] n. 度	1-1
depend on 取决于，依靠	6-3
design [dɪ'zaɪn] v. 设计	6-3
detergent [dɪ'tədʒənt] n. 洗涤剂	1-3
dice [daɪs] v. 切丁	1-1
diet ['daɪət] v. 规定饮食	4-4
dietary ['daɪətəri] adj. 饮食的	4-4
dim sum 小吃	2-2
dip [dɪp] v. 浸	4-2
disease [dɪ'ziːz] n. 疾病	1-2
dispose of sth. 处理某事	1-3
do sb. a favor 帮某人一个忙	4-1
doneness ['dʌnnɪs] n.（肉）的熟度	4-3
dough [dəʊ] n. 生面团	2-1
dredge [dredʒ] v.（用面粉、糖等）撒	4-2
dried [draɪd] adj. 干的	6-1

E

effect [ɪ'fekt] n. 效果	4-4
egg yolk 蛋黄	5-2
energetic [enə'dʒetɪk] adj. 精力充沛的	6-3
enjoy doing sth. 喜欢做某事	P-U
equal ['iːkwəl] adj. 相同的	3-1
era ['ɪərə] n. 时代	4-3
evaporate [ɪ'væpəreɪt] v.（使）蒸发	4-4
experiment [ɪk'sperɪmənt] v. 尝试	P-U

F

face up 朝上	1-1
famous ['feɪməs] adj. 著名的	4-3
fan [fæn] v. 摆成扇形	3-1
far away from... 离……很远	4-1
fat content 脂肪含量	2-1

fiber ['faibə] *n.* 纤维素		4–4
fillet ['filit] *n.* 去骨鱼片		5–2
filling ['filiŋ] *n.* 馅		2–1
fin [fin] *n.* 鱼翅		5–2
finger ['fiŋgə] *n.* 手指		1–1
flat [flæt] *adj.* 平坦的		4–2
flavor ['fleɪvə(r)] *n.* 味道		P–U
flavorful ['fleɪvəful] *adj.* 可口的		6–3
flesh [fleʃ] *n.* 果肉		3–1
flour ['flauə(r)] *n.* 面粉		2–1
flour ['flauə] *v.* 在……上撒面粉		4–1
form [fɔːm] *v.* 成形		6–1
from top to bottom 从顶部到底部		3–2
fruit platter 水果拼盘		3–1
frying pan 炸锅		4–2
further ['fəːðə] *adv.* 进一步		4–2

G

gamy ['geimi] *adj.* 有腥味的		5–1
germ [dʒəːm] *n.* 细菌		6–2
gill [dʒil] *n.* 鱼鳃		5–2
gingerbread ['dʒindʒəbred] *n.* 姜饼		3–2
give sb. the opportunity to do... 给某人做……的机会		6–3
grab [græb] *v.* 抓取		1–1
graduate ['grædʒu,eit] *v.* (从……) 毕业		P–U
grated ['greitid] *adj.* 磨碎的		4–2
grind [graind] *v.* 碾磨		6–1
gut [gʌt] *v.* 取出……的内脏		5–2

H

handle ['hændl] *v.* 处理		1–3
happen to 正巧		4–4
have a passion for... 对……有激情		2–1
heat up 加热		6–2
hoisin sauce 海鲜酱		4–3
horrible ['hɔrəbl] *adj.* 恐怖的		2–2
household ['haushəuld] *adj.* 家喻户晓的		4–3
household name 家喻户晓的名字		4–3

I

icicle ['aisikl] *n.* 冰柱		3–2
impact ['impækt] *n.* 影响		6–3
imperial [im'piriəl] *adj.* 帝国的		4–3
impression [im'preʃən] *n.* 印象		6–3
impressive [im'presiv] *adj.* 令人赞叹的		6–3
in contact with 接触		6–1
in fact 事实上		4–4
in half 成两半		3–1
in the...decade 在……10年间		6–2
incorporate [in'kɔːpəreit] *v.* 并入		5–2
incorporate into 并入		5–2
innovate ['ɪnəveɪt] *v.* 创新		P–U
involve [in'vɔlv] *v.* 包含		5–2
iron ['aiən] *n.* 铁		1–2

J

jar [dʒɑː] *n.* 罐		6–2
jerky ['dʒəːki] *n.* 肉干		6–1
juicy ['dʒuːsi] *adj.* 多汁的		4–2
julienne [dʒuːli'en] *v.* 把(食物)切成细丝		6–1

K

kitchenware store 厨具店		4–2

L

lace [leis] *v.* 系牢, 系紧		4–3
lamb [læm] *n.* 羔羊肉		5–1
land a job 找到工作		1–3
lemon juice 柠檬汁		5–2
liquid ['likwid] *n.* 液体		1–1
luncheon ['lʌntʃən] *n.* 午餐会		6–3

M

main course 主菜		6–3
major in 主修		6–3
mango ['mæŋgəu] *n.* 芒果		3–1
marketplace ['mɑːrkit'pleis] *n.* 市场		2–2

词汇	章节
material [məˈtiəriəl] n. 材料（食材）	4-2
medium [ˈmiːdiəm] adj.（牛排）五分熟	4-1
medium rare （牛排）三分熟	4-1
medium well done （牛排）七分熟	4-1
melon [ˈmelən] n. 甜瓜	3-1
menu [ˈmenjuː] n. 菜单	6-3
mince [mins] v. 切碎，剁碎，绞碎	1-3
mineral [ˈminərəl] n. 矿物质	1-2
mint [mint] n. 薄荷	3-1
mixer [ˈmiksə] n. 搅拌器，混合器	1-3
moisture [ˈmɔistʃə] n. 水分	1-3
multiple [ˈmʌltipl] adj. 多个的	1-1
muscle [ˈmʌsl] n. 肌肉	5-1
mustard [ˈmʌstəd] n. 黄芥末	1-2

N

词汇	章节
not only...but also... 不仅……而且……	P-U
nutmeg [nʌtmeg] n. 肉豆蔻	3-2
nutrition [njuːˈtriʃən] n. 营养	4-4

O

词汇	章节
on both sides 两侧都	4-3
onion [ˈʌnjən] n. 洋葱	1-1
operate...machine 操作……机器	1-3
operation [ˌɒpəˈreɪʃən] n. 运营	P-U
outer [ˈautə] adj. 外表的	5-1
oval [ˈəuvəl] adj. 椭圆形的	6-1
overcook [ˈəuvəˈkuk] v. 烘焙过度，煎焦	4-1
oversee [ˌəuvəˈsiː] v. 监督，管理	P-U

P

词汇	章节
parsley [ˈpɑːsli] n. 西芹	4-2
part-time job 兼职	5-1
pastry [ˈpeistri] n. 面点	2-1
patty [ˈpæti] n. 肉饼	6-1
peel [piːl] v. 剥皮	4-4
Peking Roast Duck 北京烤鸭	4-3
personality [pəːsəˈnæliti] n. 性格	6-3
physician [fiˈziʃən] n. 内科医生	1-2

词汇	章节
pickled meat 腌肉	1-1
pickling liquid 腌汁	1-1
pinch [pintʃ] v. 捏住	2-2
pinkish [ˈpiŋkiʃ] adj. 浅粉色的	5-1
pipe [paip] vt. 裱花	3-2
pit [pit] n. 果核	3-1
platter [ˈplætə] n. 拼盘	3-1
poach [pəutʃ] v. 水煮	5-2
powder [ˈpaudə(r)] n. 粉末	2-1
preference [ˈprefərəns] n. 爱好	4-1
preservation [ˌprezəˈveiʃən] n. 保存	6-2
preserve [priˈzɜːv] v. 保存	6-2
pressure [ˈpreʃə] n. 压力	3-2
pretzel [ˈpretsəl] n. 椒盐脆饼	3-2
prevent sth. from... 防止某物……	2-2
prize [praiz] v. 珍视	4-3
proceed [prəˈsiːd] v. 进行下去	3-1
proceed to 接着做	3-1
process [ˈprəuses] n. 过程	6-2
process [ˈprəuses] v. 处理	4-2
profession [prəˈfeʃən] n. 职业	2-1

R

词汇	章节
ramekin [ˈræmikin] n. 小盘子	3-2
rare [rɛə] adj.（牛排）一分熟	4-1
raw [rɔː] adj. 生的	4-1
recipe [ˈresəpi] n. 食谱	1-1
recorded history 有记载的历史	6-2
refer to 指代	2-1
reliable [riˈlaiəbl] adj. 可靠的	6-2
remove...from... 从……中移出……	6-2
resume [ˈrezjumei] n. 简历	6-3
reverse [riˈvəːs] v. 祛除	1-2
rinse [rins] v. 冲洗	3-2
roast [rəust] adj. 烤的	4-3
root [ruːt] n. 根	1-1
rub [rʌb] v. 揉擦	1-3
running water 自来水	1-3

S

salmon ['sæmən] *n.* 三文鱼	5-2
saute ['səutei] *v. n.* 煎，炒	4-2
saute pan 煎炒锅	4-2
savory ['seivəri] *adj.* 美味的	2-2
saw [sɔ:] *n.* 锯	5-1
scale [skeil] *v.* 去鳞	5-2
scissors ['sizəz] *n.* 剪刀	1-3
scrub [skrʌb] *v.* 擦洗	1-3
season...with... 用……给……调味	4-4
separate ['sepəreit] *v.* 分开	5-2
sesame ['sesəmi] *n.* 芝麻	2-2
set...aside 把……放在一边	3-2
shape [ʃeip] *n.* 形状	3-1
shrimp [ʃrimp] *n.* 虾肉	4-2
side by side 并排	3-2
similar ['similə] *adj.* 类似的	3-1
slice [slais] *v.* 切片	1-1
slice down 切下	3-1
slow down 减慢速度	6-2
smoke fish 做熏鱼	6-2
sound+*adj.* 听起来……	3-1
source [sɔ:s] *n.* 来源	4-4
spicy ['spaisi] *adj.* 辣的	2-2
spinach ['spinidʒ] *n.* 菠菜	1-2
spine [spain] *n.* 脊椎	5-2
spoil [spɔil] *v.* 变质	1-3
stack [stæk] *v.* 使成叠地放在……	6-1
stack up 叠放	6-1
staple ['steipl] *adj.* 主要的	2-2
start doing sth. 开始做某事	5-1
steak [steik] *n.* 牛排	6-1
stem [stem] *n.* 茎	1-1
stem end 根茎端	1-1
stew pot 炖锅	1-1
stinky ['stiŋki] *adj.* 发臭的	2-2
store...in the fridge 把……放进冰箱	3-2
strip [strip] *n.* 条状物	3-2
suggest (that) sb. (should) do... 建议某人做……	1-2
sweep the floor 扫地	5-1
sweet bean sauce 甜面酱	4-3

T

take off 取下	5-2
take on 显现	5-1
tart [tɑ:t] *n.* 蛋挞	2-1
taste wonderful 味道非常好	5-2
tender ['tendə] *adj.* 嫩的	5-1
tendon ['tendən] *n.* 筋	2-2
term [tə:m] *n.* 术语	4-1
texture ['tekstʃə] *n.* 质地	5-1
thermometer [θə'mɔmitə] *n.* 温度计	1-2
thicken ['θikən] *v.* (使)变稠	5-2
throughout [θru:'aut] *prep.* 各处	4-1
tip [tip] *n.* 技巧	3-2
toast [təust] *n.* 烤面包，吐司	1-3
tough [tʌf] *adj.* (肉)老的	4-1
tradition [trə'diʃən] *n.* 传统	P-U
treat [tri:t] *v.* 处理	6-2
turnip ['tə:nip] *n.* 萝卜	4-1
typical ['tipikəl] *adj.* 典型的	2-2

U

uncooked [ʌn'kukt] *adj.* 生的	1-3
uncooked meat 生肉	1-3
unique [ju'ni:k] *adj.* 独特的	P-U
universal [ju:ni'və:s(ə)l] *adj.* 通用的	P-U
unload [ʌn'ləud] *v.* 从……卸下货物	1-3
unmold [ʌn'məuld] *v.* 把……从模子里取出	4-1
unpack [ʌn'pæk] *v.* 从……取出	1-3
used-by date 保质期	1-3

V

vacuum ['vækjuəm] *n.* 真空	6-2
varieties of 各种	6-1
variety [və'raiəti] *n.* 种类，各种	6-1

various kinds of... 各种各样的……		2–1
vinegar ['vinigə] *n.* 醋		1–2
vitamin ['vitəmin] *n.* 维生素		1–2

W

wafer ['weifə] *n.* 威化饼		3–2
well done （牛排）全熟		4–1
what's more 除此之外		1–3
whisk [hwisk] *n.* 搅拌器		3–2
wrapper ['ræpə] *n.* 包装材料（饺子皮）		2–2

Y

yogurt ['jəugət] *n.* 酸奶		3–1
yolk [jəuk] *n.* 蛋黄		1–2

Z

zest [zest] *n.* 柑橘外皮		3–2

Appendix 4

Translation

Pre-Unit

Dialogue
S：大家早上好。我有一个重要的消息要宣布。今天有一位新成员加入我们的团队。请欢迎利奥！
T：欢迎你，利奥！
L：谢谢大家。能来到这里我真的很激动。我的名字是利奥·怀特。今年毕业于ABC学校。
S：利奥，我想让你认识一下我们团队的一些主要成员。这是我们的副总厨萨拜娜。她负责监督厨房的日常运营。
Sa：嗨，利奥。很高兴你能来。我们的厨房是我们餐厅的核心，也是奇迹发生的地方。你会发现它井井有条，效率很高。
L：谢谢您，萨拜娜。我期待向您学习。
S：接下来，你将与三厨杰基密切合作。她将会带你参观餐厅和工作场所。
J：欢迎你，利奥。你可以随时问我关于我们厨房的任何问题。
L：谢谢您，杰基。我肯定会有很多问题要问！

Reading
食物是一种将人们聚集在一起的通用语言。烹饪不仅关乎味道，还关乎文化和传统。正如炸薯条、汉堡包、比萨饼、粽子、月饼和饺子一样，每一道菜都讲述着文化和传统特有的故事。食物是联系全世界人民的最佳方式之一。

自打年幼时起，我就梦想成为一名厨师。我喜欢通过我精心准备的食物让人们感到快乐。成为一名厨师，意味着不仅要烹饪，而且要创新、尝试，并与他人分享食物带来的喜悦。

食物的世界无边无际，我的梦想也无边界。

Unit 1

● Lesson 1

Dialogue

L：贝蒂，腌肉味道不错。我想学着做一做。

B：好的。其实做法很简单。你有炖锅吗？

L：有的。这个可以吗？

B：不行，这个太小了。需要个大的。

L：好的，那这个呢？

B：非常好。我们还需要一个陶器或一个玻璃容器。

L：明白。这个就可以。

B：很好。我们现在要6磅盐、1磅糖和4加仑的水。

L：这些是6磅盐、1磅糖和4加仑的水。我们接下来要做什么？

B：现在我们将4加仑水加热至高温。好了，水在沸腾了。现在加入所有的盐和糖。

L：明白。

B：把锅从火上拿下来，让腌制配料冷却到室温。

L：然后呢？

B：然后将冷却的腌制配料倒入大陶器，将肉也放入其中。记住，用干净的砧板压在肉上，使肉能充分浸泡在腌汁中。

L：那我如何能做到这一点呢？

B：我一般用一块干净、平整且比较沉的石块。

L：真的这么简单吗？我等不及想尝一尝了。

B：不行，肉要在腌汁里浸泡3天。

L：你的意思是3天后我就能品尝到美味的腌肉了？太好了！谢谢你教我。

B：不用谢。此外，腌肉的配料汁还可以反复使用。

Reading

很多菜肴都需要配洋葱丁、洋葱片或洋葱末。只要想想你每周要切多少次洋葱，你就明白学会安全正确地切洋葱为何如此重要了。

用菜刀切去洋葱的根茎端，切的时候要留下来一点以便抓着剥皮。外皮务必要全部剥掉。

把洋葱放在砧板上，注意洋葱根端要朝上，然后把洋葱竖切成两半。通过将根部靠拢并齐，让洋葱在切的时候不易散。

把切下的半个洋葱分别平放在砧板上。从顶部开始竖切多次，根部仍要保持不断。

把洋葱旋转90度，然后横切多次，切时要注意手指保持弯曲，以免切伤。需要切多少次要视你需要丁的大小而定。

Lesson 2

Dialogue

L：赛琳，你在做什么？
C：我在做芥末酱火腿。
L：听起来味道应该不错。你怎么做的？
C：将面包屑、牛奶和鸡蛋一起放入大碗里。现在，我将肉和配料一起碾碎并充分搅拌。
L：好的。我能帮你做些什么？
C：请帮我把红糖、蒜瓣以及黄芥末放在一个小碗里搅拌，（搅好后）分撒在两个长形烤盘上。
L：好的。
C：我将搅拌好的肉糜平铺在上面。
L：明白了。
C：将肉糜烘烤一下，不要盖盖子，温度调到350华氏度，烤1小时，到肉糜温度计显示160华氏度为止。我们要等10分钟。
（10分钟后。）
C：我要放些酱汁，并和蛋黄、芥末、醋、糖、水和盐放在锅里搅拌。在低温中烹饪翻炒，直到拌料变浓稠，维持160华氏度5分钟为佳。
L：时间到。我们可以起锅了吗？
C：可以。加入黄油和辣根酱搅拌。等它冷却。包上奶油。和火腿一起上菜。

Reading

汤品中含有很多有助于你健康的重要成分——主要有维生素、矿物质。当你生病时，你可能没胃口进食。因此，你的医生可能会建议你喝汤。维生素和矿物质都对人体健康发育至关重要。它们不仅可以预防某些疾病，还可以祛除疾病。有些维生素可以帮助人体吸收矿物质。例如，维生素D帮助吸收钙，正如维生素C帮助吸收铁。乳制品汤或鱼汤中可能含有维生素D，维生素C可以在含有番茄、菠菜的汤中找到。

Lesson 3

Dialogue

L：丝黛拉，我听说你是3年前找到这份厨师助理的工作的。
S：是的。厨师助理的工作是一项要求较高的工作。
L：厨师助理有哪些职责呢？
S：我协助卸货、取出和安全储存食物。
L：你需要清洁厨房设备吗？
S：我负责清洁厨房地面、墙面、冰箱、烤箱和工作台面，清洗、操作洗碗机，手工清洗各种锅、盘、电饭煲、剪刀和其他设备，还要收拾和处理垃圾。

L:你要做烹调食物之类的事吗?

S:要的。我要洗菜、切菜,去鱼皮、切鱼片,剁肉、绞肉。我还需要做热的和冷的三明治、吐司、汤、点心、简易色拉和水果拼盘,准备茶和咖啡。

L:嗯,确实很具挑战性。

S:除此之外,我得会使用电动搅拌器、削片机、刀具、切割工具,调拨存货和查看保质期。

Reading

如果你想食用没有与虫子、化学药品一起出现的水果和蔬菜,要在吃之前把它们清洗干净。

首先要保证你的工作台面、冰箱和餐具都干干净净。

做饭或处理蔬果之前请洗手。

分开放置新鲜蔬果和生肉。

需要用到蔬果时再清洗。蔬果都有天然的保护层来锁住自身水分。提前清洗容易导致变质。

在自来水下轻柔地揉擦水果和蔬菜。忌用任何肥皂或化学洗涤剂。

对于比较硬的蔬果,如苹果和土豆,在用清水清洗的同时可用蔬菜刷用力擦洗来除去污垢。

Appendix 4
Translation

Unit 2

Lesson 1

Dialogue

I：可以谈谈您面点师的职业生涯吗？
B：我8岁就开始烹饪。那时候，我从没把烹饪当成一种职业。
I：那您什么时候开始有了这样的想法呢？
B：我上烹饪学校的时候。第一次上烘焙课之后，我就意识到这是我真正想做的事。
I：对那些对面点师职业感兴趣的人，您有什么建议吗？
B：好好思考自己是否真的想做这个，确信你对这份职业是有激情的。
I：您为什么喜欢这份工作？
B：作为面点师，我可以掌控我的事业。我有业余时间尝试新食谱。
I：什么因素会影响面点师的收入？
B：主要因素是作为厨师和面点师的技术水平、经验和名声。学习糕点装饰技能可以提高面点师的薪水。

Reading

面点是指用面粉、牛奶、黄油、发酵粉和鸡蛋等原料做成的各种各样的烘烤食品。小蛋糕、果馅饼、蛋挞和其他烘烤甜点都可以称作面点。面点还可以指代制作这些烘烤食品的生面团。

面点和面包的区别在于面点内的脂肪含量更高。好的面点质地轻软，但足以稳固支撑馅料重量。其他面点如丹麦酥皮甜饼、羊角面包，其特有的多层酥皮质地是通过反复压擀生面团，在其上涂抹黄油，将其折叠成多个薄层而获得的。

Lesson 2

Dialogue

L：特蕾西，我想学包饺子。你能指点指点我吗？
T：我的荣幸。首先，你得把所有的食材都准备好。1杯水、2茶勺盐、1磅瘦肉糜、2大汤勺酱油、1大汤勺芝麻油、1茶勺生姜、1大汤勺葱、1茶勺黑胡椒、1茶勺白胡椒和2大勺米酒。
L：明白。我怎么开始做呢？
T：很简单。你把所有食材放在一起，然后搅拌。
L：好的。但我怎么做饺子皮呢？
T：你可以在超市买饺子皮。把饺子皮放在手上，舀一小勺馅料放在饺子皮中间。将饺子皮对折，让它看起来像个半圆。将皮重叠的两边捏住。然后，从饺子一端朝另一端折一些小褶子。

131

L：我需要做的准备工作就是这些吗？

T：还不是。最后一步是在锅里的沸水中煮饺子。轻轻搅动几下防止饺子粘在一起，然后就可以起锅了。

Reading

中式小吃有着悠久的历史，在中国的聚集场所、车站、市场、繁华街头以及如纽约市中国城等地方都可以方便地买到。下面介绍一些典型的中式小吃。

包子

在中国，包子是一种主食，可在早餐或一天中的任何时间食用。通常在蒸包子前，人们会在面团里加些美味的馅料。

凤爪

凤爪基本上是由皮和筋构成的，所以骨小、质地耐嚼。一般将其放入辣椒酱腌制。

臭豆腐

臭豆腐是以味臭而著名的中式小吃。臭豆腐常作为沿街小吃。

Unit 3

Lesson 1

Dialogue

L：温迪，你能教我做冷汤吗？
W：可以。教你做黄瓜汤如何？
L：听起来不错。
W：你需要1根黄瓜、14片新鲜薄荷叶和2杯白脱牛奶。
L：我需要怎么处理黄瓜？
W：制成黄瓜丝。你还要准备2大汤勺酸奶、2大汤勺橄榄油、盐和胡椒。
L：然后我能做什么？
W：留下4片薄荷叶摆盘。将剩下的薄荷叶剁碎，和黄瓜丝一起放入食物搅拌机。搅拌到均匀为止。
L：明白。我会继续搅拌，再加入白脱牛奶、酸奶和橄榄油，搅拌成均匀的液体。
W：别忘了加胡椒和盐。
L：最后，用4片薄荷叶装饰。
W：没错。

Reading

看上去很专业的水果拼盘比较容易做。关键是将水果大小、形状切得类似。在这个拼盘里，我选用的是甜瓜、草莓和芒果。

为在拼盘中心摆出心形，要去除草莓的绿叶，从底部朝上切（顶端不切断），并将其摆成扇形。

开始将草莓放入盘中组成心形。

继续切其他水果。芒果去皮，然后将果肉绕中心果核切下。切下的果肉应形状相同。将甜瓜切成两半，去籽，去皮。接着切成大小相同的果块。

将水果围绕中间的心形摆开。将相同水果的果肉放在一起，在拼盘上摆成扇形。

Lesson 2

Dialogue

L：梅，你能教我做柠檬酸奶奶油吗？
M：可以。你得准备原味浓稠型酸奶、4个鸡蛋、100克黄油、2个柠檬、110克糖、1大汤勺蜂蜜、4片姜饼、2撮肉豆蔻。
L：好的。怎么开始做呢？

M：把柠檬洗净。去皮，切成条状。放入凉水中。煮至沸腾3分钟。在自来水下冲洗并把水沥干。将2勺水与30克糖一起加热。同时在糖汁中加入柑橘外皮，低温加热8~10分钟。将其放在盘里。

L：明白。下一步是不是切4片姜饼，并把姜饼放在底部？

M：是的，将蜂蜜和2撮肉豆蔻放入酸奶中搅拌，分别放到一个个小烤盘里，然后放进冰箱1小时。你能告诉我接下来做什么吗？

L：榨柠檬。

M：对。将柠檬汁、鸡蛋、剩下的糖放在碗里搅拌，将碗放入双层蒸锅里，用电动搅拌器一直搅拌到食物凝固、泡沫丰富为止。将其从双层蒸锅中拿出。加入几块黄油。然后晾凉。

L：我知道最后一步是将柠檬奶油倒在酸奶上。将表面弄平整，放入冰箱3~4小时。

Reading

为你的朋友制作一些创造性的甜点当作礼物是一个不错的想法。下面是一些关于如何把一块普通的姜饼装饰成可爱姜饼屋的技巧。

（冰柱）用圆形挤花嘴。用先重后轻的压力从顶部到底部裱花。

（篱笆）将椒盐脆饼棒竖着并排放好。挤花袋里放入糖衣，用挤花嘴挤出糖衣粘上扶手。将其牢牢地粘好。

（壁炉/烟囱）用圆形挤花嘴画出壁炉或烟囱的外形。

（窗户）用圆形挤花嘴画出窗户的外形。用糖衣将威化饼干粘在窗户上。

Unit 4

Lesson 1

Dialogue

L：海伦，可以帮我一个忙吗？

H：好，你想让我帮你什么？

L：我想为素食者做道创意菜。你能给些建议吗？

H：蔬菜饼怎么样？

L：听起来挺有意思的。需要什么食材吗？

H：一些胡萝卜、萝卜、豌豆和豆角。

L：这些蔬菜很有营养。

H：是的。沥干水后切蔬菜。

L：好的。

H：用干巾将蔬菜包裹拍干。

L：这简单。

H：在蔬菜上撒面粉。

L：然后烘焙？

H：不是，离那步还很远。拿个碗，将澄清黄油和鸡蛋搅拌在一起。加入面粉、发酵粉、柑橘外皮、肉桂皮和调味料。

L：哦，我说错了。现在可以烘焙了，对吗？

H：是的。调到 250 摄氏度，烤 10 分钟，然后调低温度至 170 摄氏度，再烤 45 分钟。烤好后，把食物从模子里取出，切片趁热吃。

Reading

牛排是块状牛肉。

烹制牛排的时间依个人喜好而定；时短则汁多，时长则肉干而老。下列术语依次描述从生到熟的牛排：

1. 全生：完全未经烹煮的生牛肉。
2. 近生和极生：快速煎炸，里面为血红色，几乎只是温热。
3. 一分熟：(52 ℃) 表面呈灰褐色，里面为血红色，稍微温热。
4. 三分熟：(55 ℃) 牛排呈血红色，里面也煎得温热。这是大多数牛排店的标准生熟程度。
5. 五分熟：(60 ℃) 剖面为粉红色，核心仍有血红色，外围呈灰褐色。
6. 七分熟：(65 ℃) 核心为淡粉红色。
7. 全熟：(核心温度高于 71 ℃) 各处完全为灰褐色，表面稍微烤焦。

8. 煎焦：(远高于 71 ℃) 各处完全为黑焦色，味稍苦。

Lesson 2

Dialogue

S：嗨，我叫迪克·费希尔。欢迎光临我们的厨具店。
L：我在找锅。
S：您需要什么样的锅，是煎炒锅还是炸锅？
L：有什么不同吗？
S：它们都有平坦的底部，但煎炒锅有盖子，边缘较垂直，这样翻炒时食物不易弄到锅外。
L：我需要一个煎炒锅。

A：利奥，我今天教你做一种法式料理。
L：那是什么呢？
A：你拿块肉，涂上调料。撒上面粉。然后浸入搅拌好的鸡蛋、切碎的西芹和磨碎的干酪里。
L：然后呢？
A：然后你加入黄油煎炒肉块。
L：我还是第一次听说。它的味道如何？
A：味道好极了。这道菜很受法国年轻人欢迎。

Reading

有人说这是上海最好的一道菜。它的中文名叫水晶虾仁。虽然这道菜看起来是用虾仁简易制成的，但食材处理和烹制的过程相当复杂。这道菜的精华在于一入口尝到的是虾外皮的酥脆感，再进一步品尝会发现虾仁柔嫩多汁。要达到这般效果就要看厨师在烹饪时对火候和时间的掌控技巧。这道菜在上海许多餐厅都能品尝到。

Lesson 3

Dialogue

L：朵拉，你能教我怎么做烤鸭吗？
D：如果你要做烤鸭，你就得准备鸭胸肉、大蒜瓣、盐和胡椒。
L：好的。
D：在较肥的一面，用刀划上几条平行划痕。两侧都撒上盐和胡椒。
L：我知道。这么做是让味道渗进鸭肉里。
D：是的。你要在脂肪和瘦肉中间塞入一些新鲜大蒜，然后把肉固定系紧。
L：好的。我把它们放在哪里？
D：把它们放在烤盘里。
L：我怎么烤呢？

D：把温度调到 250 摄氏度，烤 15 分钟。到第 7 分钟时，除去烤盘上的脂肪，并加入一杯水。
L：也就是说，烤到一半的时候我要除去脂肪然后加水。
D：是的。烤完后，去掉鸭胸肉上的绳子，一片片切好。它的熟度得是嫩的。
L：今晚我就试一试。

Reading

北京烤鸭是北京的一道名菜，始于帝国时代，现已被认为是中国的国菜之一。

烤鸭因其皮薄、香脆以及厨师在食客面前将鸭肉切成薄片而受到珍视。鸭子在放入焖炉或挂炉烘烤之前涂上调料。鸭肉可与薄饼、葱、海鲜酱或甜面酱一道食用。北京最著名的供应烤鸭的两家店便是全聚德和便宜坊，这两家店都已有上百年的历史，现已成为家喻户晓的名字。

Lesson 4

Dialogue

L：苏珊，你能教我做番茄碎吗？
S：番茄碎是以蔬菜为主的酱料，营养非常丰富。番茄吃得越多，患癌风险越低。
L：这蔬菜真不错！
S：如果你想做番茄碎，你就要准备新鲜番茄、洋葱、大蒜瓣、橄榄油、盐、胡椒和糖。
L：好的。我知道首先要做的是给洋葱剥皮然后切好。
S：是的。然后在大锅里将橄榄油加热。加入洋葱，等待几分钟让其渗出汁水，注意不要煎焦。
L：好的。然后我要做什么？
S：加入去皮、去籽并压碎的番茄。用盐、胡椒、大蒜和少量糖给它们调味。
L：好的。这样就行了吗？
S：中火烹饪 30 分钟。烹饪时间取决于番茄的情况。最后，要使水蒸发完。

Reading

在众多食物中，豆类几乎是完美的食物。事实上，豆类不但营养丰富，而且低卡路里。若你正巧在节食，豆类正是适合你的食品。豆类是纤维素的最佳来源之一。研究表明，纤维素有助于我们预防包括心脏病和癌症在内的许多疾病。

黑豆是最有名的豆类之一。有许多烹调方法适合黑豆。黑豆是膳食纤维的一个来源。黑豆和其他豆类中的膳食纤维有控制大餐后体内血糖水平的效果。

Unit 5

● Lesson 1

Dialogue

L：肉类处理工主要做什么工作，杰森？

J：通常情况下，肉类处理工负责剔骨和切肉。

L：你是怎么成为一名肉类处理工的？

J：我基本上是自学的。我开始在屠宰场工作，以此为起点……一直负责去牛骨近20年，然后又去肉店做一些兼职，边做边学。

L：这份工作有危险吗？

J：是的，当你用锯骨机和绞肉机……你得非常小心地操作这些机器，因为一不留神，就会切到自己的手。

L：有没有成为一名肉类处理工的建议？

J：如果是学徒，别指望一下就能开店。做任何事都一样，你要从底层做起，扫地、清理碎骨，边做边学。但学徒期满，你就准备好具备从事这份工作的条件了。

Reading

肉分两种：老的和嫩的。老的肉包含肌肉，需要炖或煮来使其变嫩。而嫩的肉只要稍许烹制便可，这样可保持其原有质地和味道。

选择优质牛肉

要选外表肥肉较少的牛肉。肉质须紧实、纹路清晰并显淡樱桃红色。

选择优质猪肉

选颜色偏粉白到粉红之间、手感结实的猪肉为佳。

选择优质羊肉

选择屠宰时为5～7个月或更小的羊的羔羊肉。这种肉味道、肉质比普通羊肉鲜嫩。普通羊肉通常有腥味。

● Lesson 2

Dialogue

L：荷兰酸酱拌水煮三文鱼味道非常好。艾达，你是怎么做的？

A：谢谢。这并不复杂。首先，你准备3大汤勺新鲜柠檬汁、1茶勺橄榄油、1杯黄油、2片去皮去骨三文鱼片、3个蛋黄、盐和胡椒。

L：好的。接下来做什么？

A：找一个平底锅，把柠檬汁和橄榄油倒进去，加足够的水。用盐和胡椒给水调味，然后加入三文鱼。

L：你什么时候加三文鱼？

A：在中高温时加入三文鱼，继续加热至水烫且有水蒸气出现。

L：蛋黄有什么用？

A：把蛋黄放入金属碗，在热水中搅拌。将碗放在沸水上方，不要接触到沸水。一直搅拌到蛋黄变稠为止。

L：什么时候放黄油？

A：蛋黄变稠后，加黄油搅拌，一次一块，直到融化变成荷兰酱。接着关火，加入柠檬汁搅拌，放入盐和胡椒调味。

L：听起来挺简单的。

A：上菜前，把水煮三文鱼多余的水分沥干，逐片放在一个餐盘上，并将荷兰酱浇在三文鱼上。

Reading

切鱼包含去鳞、切下鱼翅、取下鱼鳃、取出鱼内脏并清洗干净。如菜谱要求以完整鱼形烹制，那么应沿着脊椎切。然后与脊椎平行切下直至头部，将鱼肉自头尾从中间向两侧分开。抬高中间骨头和鱼刺，在头部和尾部剪下脊椎。最后，将腹腔和外部洗净，尽可能将鱼整理成和原先一样的形状。

Unit 6

Lesson 1

Dialogue

L：杰基，你能教我切蔬菜吗？

J：不同蔬菜切法不同。

L：土豆怎么切？

J：土豆是种椭圆形的蔬菜。所以首先我们得把它切成两半，这样就有平整的一面。从两半土豆切成许多半圆形的切面。这样我们就可以直接用这些土豆片了。

L：如果我想把土豆切成细丝，我们接下来该怎么做？

J：把土豆片叠放，刀向下切，刀尖接触切面，这样就能切成细丝了。

L：那大白菜怎么切？

J：那是带叶子的菜。我们从把最大片叶子放在底部开始，使菜叶叠放，把菜叶卷紧再切。

L：非常好，那胡萝卜呢？

J：那是长形的蔬菜。因为它太长了，我们可以把它切成2~3段。每段胡萝卜切下一片薄片，这样就可以放平了。然后再开始切片，越薄越好。

L：好的。谢谢。我从你这儿学到了很多。

Reading

肉类有多种烹调法，如牛排或干肉类像牛肉干。还可以碾磨成肉饼状（如汉堡）。还有些可加工处理成熏肉，用盐或盐水腌制。其他方法还包括卤制、烧烤，或简易煮、烤、煎。肉类一般是煮熟吃，不过也有吃生肉的传统烹调食谱。

肉类是做三明治的一种典型底料。三明治中受欢迎的肉类包含火腿、猪肉、其他香肠和牛肉，如牛排、烤牛肉。肉类还可以制成罐头。

Lesson 2

Dialogue

L：萨拜娜，你能教我怎么保存食品吗？

S：好啊，看，我正在熏鱼。熏烤是保存食品的一种方法。

L：你好熟练。你怎么熏鱼呢？

S：我已经买了一些速冻的鱼。现在解冻并清理，去头、去尾、去鱼鳍，在清水里洗一下。

L：好的。那你现在在做什么？

S：我现在在准备盐水，然后我会把鱼放在盐水里15分钟。

（15分钟后）

S：现在我们得把鱼从盐水中捞出来，用冷水清洗。

L：我帮你把鱼放在涂油的烤架上。我看我妈妈总这么做。

S：谢谢。所以我们只要保证温度不过高，大约在65摄氏度，2个小时后，把温度上升至95摄氏度。

L：挺简单的，我们一直熏到鱼的鱼皮易剥落并全熟，是吗？

S：聪明！

Reading

食品保存是处理食品以阻止或大幅度减慢其腐烂速度的过程。保存的方法包括烘干、冷冻干燥、真空包装和装罐保存。

保存果蔬食品的传统做法是装罐保存。这包括将罐子和产品加热，杀灭细菌，并将产品密封。

最可靠的古老保存方法是风干。在有记载的历史之前，人们就风干草本植物、肉类、蔬果备用。

近10年来有一种新型、流行的保存食品的方法。真空包装是将食品放入一种厚塑料袋，将袋内空气抽干密封的方法。

Lesson 3

Dialogue

H：早上好，史密斯先生。我叫霍华德，利奥的朋友。

S：早上好，霍华德。请坐。利奥向我们推荐你，他对你评价很高。你申请了我们餐厅的厨师的岗位。你能先介绍一下自己吗？

H：好的。谢谢您给我这次面试的机会。我从ABC学校毕业，主修烹饪艺术。我曾获华东区职业院校学生烹饪比赛一等奖。

S：非常令人赞叹的简历。你有没有上过专业英语课？

H：上过。在学校的第2年我上过烹饪英语，并且暑假期间在一家名叫苹果蜂的美式餐厅里做过兼职。

S：你能描述一下自己的性格吗？

H：好的。我是个很好相处的人，喜欢团队合作。我非常耐心、精力充沛。我能勤奋工作且学东西很快。

S：很好。那你有什么缺点吗？

H：嗯，我没做过全职工作，我需要积累更多的经验。

S：好的。就这样。谢谢你抽空过来，霍华德。如果通过面试，你会进行烹饪考试。我们会尽早通知你。

H：非常感谢史密斯先生。再见。

Reading

在设计菜单时，你应该设计前后和谐融合的几道菜。记得设想影响和最终印象。

例如，如果你要举办一个以意大利香肠芥末为主菜的午餐会，前后菜最好是清淡的，如简易色拉和多汁可口的水果。这可以清胃和防止食欲过盛。

开胃品：如只提供开胃品，那一般是每位成年人6～8块。如在正餐或自助餐服务之前上开胃品，4～6块足矣。更可取的是品种少但数量多。这样可以节约时间和成本。开胃品的量还可以取决于其种类。